# Biological Aspects of Human Sexuality

## second edition

**HERANT A. KATCHADOURIAN, M.D.**
**DONALD T. LUNDE, M.D.**
*Stanford University*

**HOLT, RINEHART AND WINSTON**
*New York   Chicago   San Francisco*
*Dallas   Montreal   Toronto   London   Sydney*

Cover illustration, Leonardo da Vinci, figures in coition

**Library of Congress Cataloging in Publication Data**

Katchadourian, Herant A.
  Biological aspects of human sexuality.
  Bibliography: p. 195
  Includes indexes.
  1. Human reproduction.  2.-Sex.  1. Lunde,
Donald T., joint author.  II. Title.
QP251.K36      612.6      80-11649
**ISBN 0-03-055396-2**

# Preface to the Second Edition

In the five years since the publication of the first edition, *Biological Aspects of Human Sexuality* has found wide use in human sexuality courses in the United States and Canada. During this period the authors and the publisher have received many letters of comment from instructors who have taught from the book. While these comments generally have been favorable, they have contained useful suggestions for revision. Many of the changes made in the second edition stem from these suggestions.

All chapters have been substantially brought up to date, including reference to the most recent research findings.

Recognizing the difficulty that male writers have in reflecting the woman's as well as the man's experience and perception of human sexuality, particular heed has been paid to the suggestions of women reviewers. Their comments have been an invaluable aid in balancing the picture.

A new feature in this edition is the addition of boxed material throughout. The boxed material is designed to highlight particularly interesting information, to provide expanded documentation, and to amplify points in the text.

The division of labor between the co-authors was as follows: Katchadourian wrote the introductory chapter, the chapters on anatomy, and on physiology. Lunde was responsible for chapters on hormones, conception and pregnancy, contraception, and diseases of the sex organs. In the revision of his chapters Lunde received invaluable assistance from John and Janice Baldwin of the University of California, Santa Barbara. The revised chapters were planned by Lunde and the Baldwins, the Baldwins revised the chapters, and then, on the basis of Lunde's critiques prepared final drafts. The Baldwins, experienced in teaching the human sexuality course, bring a fresh perspective to the chapters that they prepared.

January 1980                                    The Publisher

# Contents

# Introduction

Courses in human sexuality are a relatively recent phenomenon. Although such courses are expanding rapidly across the country, the teaching of human sexuality remains quite undifferentiated as a separate field. There are as yet no departments or probably even full-time teaching positions in this field. An institution will have such a course if there is an instructor willing to teach it, and these who currently do so include biologists, psychologists, anthropologists, physicians in various specialties, and others.

As a result textbooks in human sexuality are confronted with the task of having to be all things to all persons. This complicates the task, but it is also a blessing in disguise. It forces authors to take a multidisciplinary approach, which at a time of increasing specialization allows an opportunity to attempt an integration of diverse sources of information.

In our larger textbook, *Fundamentals of Human Sexuality,* we took the broadest view possible and at the same time practical. In addition to the biological and behavioral perspectives, we ventured into matters of sex and society and law and morality. In our experience this greatly enhanced our efforts to provide students with an enriched and balanced view of human sexuality.

The appearance of this volume is in no way a relinquishing of our commitment to the broader perspective. Biology is the bedrock on which sex is based, but there is more to human sexuality than biology. Rather, this volume is a concession to practical but important considerations.

Current courses in this field vary widely as to scope and depth of coverage. The needs of institutions are quite different in this regard. The diversity of professional backgrounds among instructors is also such that they feel variously qualified to rely on their own expertise in some areas and depend more on the text in others. Another by no means trivial consideration is the matter of cost. Larger, profusely illustrated books are inevitably expensive.

All of these factors combined suggested the need for a shorter text. We therefore decided to issue part of the larger *Fundamentals of Human Sexuality* as a separate volume. We chose the section on biology for several reasons. First, biology would seem to constitute the most common core among such courses. Whatever part of the subject matter one teaches or chooses to omit, it is hard to conceive that anatomy, physiology, conception, and contraception could be left out. Second, the biological data are less euqivocal and do not lend themselves, as much as the behavioral information, to the personal predilections of the instructor presenting the material. Third, for these and possibly other reasons we were informed by potential users of a shorter volume that biology is the subject matter they needed most.

The chapters of this volume correspond to Chapters 2 to 7 of the third edition of *Fundamentals of Human Sexuality*. The final portion of the last chapter in this book is not part of *Fundamentals of Human Sexuality* and was written to provide an appropriate closure for this volume.

# Chapter 1

## Anatomy of the Sex Organs

Praise be given to God, who has placed man's greatest pleasure in the natural parts of woman, and has destined the natural parts of man to afford the greatest enjoyment to woman.

—The Perfumed Garden

The human body has no other parts as fascinating as the sexual organs. Venerated and vilified, concealed and exhibited, the human genitals have elicited a multitude of varied responses. They have been portrayed in every art form, praised and damned in poetry and prose, worshipped with religious fervor, and mutilated in insane frenzy.

Many of us combine a lively interest in the sex organs with an equally compelling tendency either to deny such interest or to be ashamed of it. There are men and women who have been married for years, who have engaged in sexual intercourse countless times, but who have never looked frankly and searchingly at each other's genitals. Nor is this aversion merely a matter of prudishness. To many people the sex organs appear neither beautiful nor sexy when viewed directly. Unfortunately, although concealment may promote desire, it also perpetuates ignorance.

The performance of basic procreative functions obviously does not require any formal knowledge of anatomy. For most people, however, some knowledge of sexual anatomy, particularly of the genitals, is helpful in understanding sexual functions. Anatomy, the study of structures, is related to physiology, which deals with functions, as geography is related to history: It is the description of the theater where the action takes place.

The genitalia are part of the reproductive system, and procreation is their most fundamental function. But humans also engage in sex for many other purposes. We cannot procreate without sex, but we do not always engage in sex in order to procreate. This separability of sex and reproduction is an issue of far-reaching psychological and social consequence, one still unresolved in most human societies.

An objective approach to anatomy would

deal with each bodily structure in its own right, and anatomists do just that. But most of us are likely to approach parts of the body with far more subjective attitudes. We find some organs more fascinating or think of them as more important than others, sometimes without adequate physiological justification for such views.

To think of the heart as the seat of emotions is a harmless fiction. But our attitudes towards the genital organs have not been so innocuous. Male perspectives on sexuality that have been socially dominant have held an ambivalent view of the female genitals. Men are fascinated, aroused, and preoccupied by them, but simultaneously they are feared and deprecated as defective copies of the male. Males often have ambivalent views about their own genitals as well, alternately imbuing them with exaggerated importance and being anxious about their size, shape, and capacity for performance.

Female attitudes towards the genitalia, be it their own or those of males, have had much less occasion for public expression, but women too seem to have feelings of shame, confusion, or ambivalence. Women also tend to be even less knowledgeable than men about sexual anatomy because their own organs are more concealed and they have been expected or assumed not to be interested in the genitals of males. These attitudes have recently become more open and accepting, but there is still much variability in people's views on the subject.

## The Basic Plan of the Reproductive System

There are obvious and concealed differences between the reproductive tracts of the two sexes. But before we dwell on these, it is important to understand the fundamentally similar plan on which the two are built.

In both sexes the reproductive system can be thought of as consisting of three parts. First are a pair of gonads or reproductive glands (ovaries in the female; testes in the male) that produce the germ cells (ova, or eggs, and sperm). Second are a set of tubes (a pair of fallopian tubes and the uterus in the female; epididymis, vas deferens, ejaculatory duct—in pairs—and the urethra in the male) that transport these germ cells. Third are organs for the delivery and reception of the ejaculate containing sperm (penis and vagina). In addition, the gonads produce hormones in both sexes and the uterus houses the growing embryo. There are also a number of accessory sex organs (a pair of bulbourethral glands in both sexes and a pair of seminal vesicles and the prostate gland in the male) and covering structures that form part of the genitalia (scrotal sac in the male, major and minor lips in the female).

The reproductive systems of both sexes have the same embryological origin. In fact, very early in life it is not possible to tell them apart by examining the immature reproductive structures. Although the sex of the infant is decided at the moment of fertilization, this does not become manifest until later in development. The differentiation of the male from the female is dependent on the genetic contribution of the fertilizing sperm (the Y chromosome) and on hormones produced by the developing testes (androgens). Although normally these events progress quite predictably, anatomical maleness and femaleness are not immutably fixed by genetics. It is possible to push development toward the female or male side despite the fundamental genetic sex of the person. In abnormal conditions like hermaphroditism, this is precisely what happens. We shall have occasion to discuss these matters more fully in later chapters.

From this common embryonal origin the reproductive organs differentiate in the two sexes to fulfill their complementary functions. Sexual organs can thus be compared usefully in two ways. One is in terms of their embryonal origins. Every reproductive organ in one sex has its developmental counterpart or *homologue* in the other.[1] These organs can also be compared as functional counterparts. For example, the clitoris and penis are homologous. Functionally, they are also comparable by being highly responsive to erotic stimulation. But the clitoris has no direct reproductive function. It does not mate with the penis; that is done by the vagina, which is thus the counterpart of the penis in coital terms.

One could conclude from what has been said so far that we seem to consider heterosexual intercourse for reproductive purposes the standard or normative unit for sexual functions. This is and is not the case, depending on what one means. The anatomy and physiology of the reproductive system has evolved in ways to maximize its reproductive potential. Its design, as it were, would therefore make most sense in reproductive terms. But this is not to say that the only or even best use of sex for a given individual is necessarily for procreation. But even if sex is not to be used for reproduction, the structure and function of the sex organs can be best understood within that biologically fundamental framework.

We shall now turn to the main purpose of this chapter by examining first the bony framework which houses the reproductive system. Then we shall deal with the anatomy of the female and male sex organs and close the chapter with a consideration of the developmental processes. The reason that we shall consider the female first is not because she is more or less "important" than the male, but because the female reproductive system is somewhat less convoluted than that of the male and thus probably easier to understand for most.

[1]In biology the term "homologous" refers to organs or parts that correspond to each other in evolutionary origin and are basically similar in structure but not necessarily in function. By contrast, "analogous" organs or parts resemble functionally but have different origins or structure.

## The Bony Pelvis

The reproductive system in either sex is located partly inside the body cavity and partly outside it. Although all the sex organs belong to a single system, the internal ones are regarded as primarily organs of procreation, whereas the external ones, the genitals, are associated more closely with sexual activity itself. The external organs are thus more likely objects of erotic and social interest.

The internal sex organs are housed in the *pelvis* (*see* Figure 2.1). The bones of the pelvis consist of the triangular end of the vertebral column (*sacrum*) and a pair of "hip bones," which are attached to the sacrum behind and to each other in front (at the *symphysis pubis*), thus forming a circle at their rim. Each "hip bone" actually consists of three separate bones (*ilium, ischium, pubis*) that are fused together. The components of the bony pelvis are in turn fixed and permit no movement.

Comparison of the female pelvis with that of the male in Figure 2.1 shows that even though the male pelvis has a heavier bone structure, the female pelvis is broader and its outlets wider. This allows for easier passage of the infant's head during birth. Women with narrower or male-type pelvic outlets sometimes cannot give birth naturally, and their children must be delivered through abdominal (cesarean) section of the uterus (discussed in Chapter 5).

The pelvic cavity is a bottomless basin crowded with organs belonging to the reproductive, urinary, and digestive systems. Figure 2.2 and Figure 2.3, respectively, show the relationship of the female and male reproductive organs to the pelvis.

Separating these organs from one another and supporting and attaching them to the bony framework of the body are various tough fibrous structures analogous to canvas sheets (*fasciae*) and cords (*ligaments*). These structures, along with the various muscles in the area, form a multilayered "hammock" in

FEMALE                    MALE

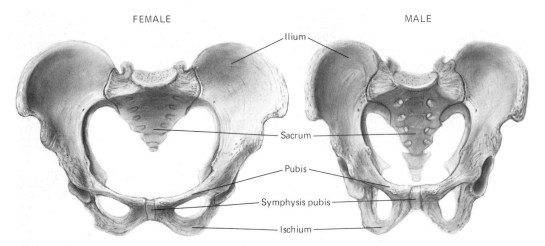

Ilium

Sacrum

Pubis

Symphysis pubis

Ischium

**Figure 2.1** The bony pelvis. From Dienhart, *Basic Human Anatomy and Physiology*. Philadelphia: Saunders, 1967, p. 35. Reprinted by permission.

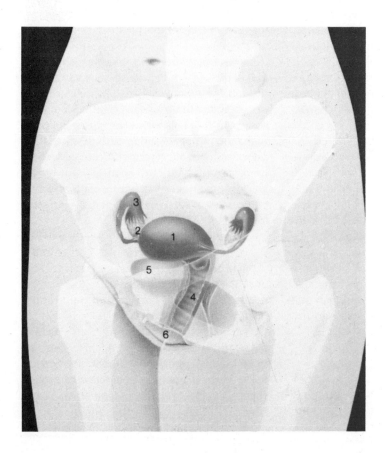

**Figure 2.2** Female reproductive organs. (1) Uterus, (2) ovary, (3) Fallopian tube, (4) vagina, (5) bladder, (6) labia majora and labia minora on the right side. From Nilsson, *A Child Is Born*. New York: Delacorte Press, 1977, p. 21. Reprinted by permission.

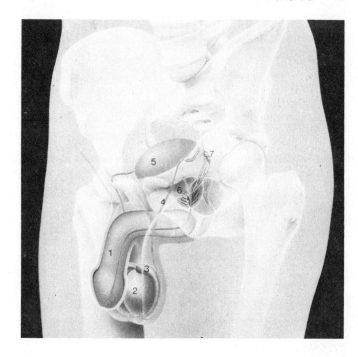

**Figure 2.3** *Male reproductive organs.* (1) Penis, (2) testicle, (3) epididymis, (4) spermatic cord, (5) bladder, (6) prostate, (7) seminal vesicle. From Nilsson, *A Child Is Born.* New York: Delacorte Press, 1977, p. 17. Reprinted by permission.

which the genital organs are embedded and suspended.

Because of the anatomical peculiarities of the reproductive organs in the two sexes, the particular arrangements of these supporting fascia, ligaments, and muscles are somewhat different, as will become clear when these structures are described.

## Female Sex Organs

### External Genitals

The external genitals of the female are collectively called the *vulva* (''covering''). They include the *mons pubis* (or *mons veneris,* ''mount of Venus''), the *major* and *minor lips,* the *clitoris,* and the *vaginal introitus,* or opening.

The mons pubis is the soft, rounded elevation of fatty tissue over the pubic symphysis. After it becomes covered with hair during puberty the mons is the most visible part of the female genitals.

### The Major Lips

The major lips (*labia majora*) are two elongated folds of skin that run down and back from the mons pubis. Their appearance varies a great deal: Some are flat and hardly visible behind thick pubic hair; others bulge prominently. Ordinarily they are close together, and the female genitals appear ''closed.''

The major lips are more distinct in front. Toward the anus they flatten out and merge with the surrounding tissues. The outer surfaces of the major lips are covered with skin of a darker color, which grows hair at puberty. Their inner surfaces are smooth and hairless. Within these folds of skin are bundles of smooth muscle fibers, nerves, and vessels for blood and lymph. The space between the major lips is the *pudendal cleft*;[2] it becomes

---

[2]*Pudendum,* a term no longer in general use, refers to the female external genitalia and means ''a thing of shame.''

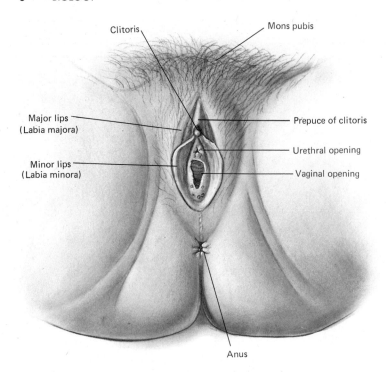

Clitoris — Mons pubis

Major lips (Labia majora) — Prepuce of clitoris

— Urethral opening

Minor lips (Labia minora) — Vaginal opening

Anus

**Figure 2.4**  External female genitalia. From Dienhart, *Basic Human Anatomy and Physiology.* Philadelphia: Saunders, 1967, p. 217. Reprinted by permission.

visible only when the lips are parted (see Figure 2.4).

### The Minor Lips

The minor lips (*labia minora*) are two lighter-colored hairless folds of skin located between the major lips. The space that they enclose is the vaginal *vestibule* into which open the vaginal and urethral orifices, as well as the ducts of the Bartholin's or greater vestibular glands. The minor lips merge with the major lips behind. In front each divides in two: The upper portions form a single fold of skin over the clitoris and are called the *prepuce of the clitoris;* the lower portions meet beneath the clitoris as a separate fold of skin called the *frenulum of the clitoris.*

From front to back the structures enclosed by the minor lips are the clitoris, the external urethral orifice, the vaginal orifice, and the openings of the two greater vestibular glands. The anus, which is completely separate from the external genitals, lies farther back.

### The Clitoris

The clitoris ("that which is closed in") consists of two masses of erectile spongy tissue (*corpora cavernosa*), the tips (crura) of which are attached to the pubic bone. Most of the clitoral body is covered by the upper folds of the minor lips, but its free, rounded tip (the glans) projects beyond it. The urethra does not pass through it.

The clitoris becomes engorged with blood during sexual excitement. Because of the way it is attached, however, it does not become erect as the penis does. Functionally it corresponds more closely to the glans of the penis: It is richly endowed with nerves, highly sensitive, and a very important focus of sexual stimulation, which is its sole function.[3] The clitoris has also been subjected to ritual mutilation (*see* Box 2.1).

[3]Given its importance to female sexual arousal, the clitoris has recently become the focus of more professional attention. For a detailed study of clitoral anatomy, function and related considerations, see Lowry and Lowry (1976). Also of interest is Lowry (1978).

**Box 2.1**   Female "Circumcision"

The practice of female "circumcision" is far less known generally than its male counterpart. Yet the practice has been widespread in many cultures and continues to be practiced in the Near East and the African continent where currently there are an estimated 30 million women who have undergone one or another version of this mutilating procedure.[1]

To be precise, the term female "circumcision" should be restricted to the removal of the prepuce of the clitoris only, which is the common practice among Moslems. Other procedures include the amputation of the clitoris (*clitoridectomy*), sometimes with the additional removal of the labia minor and the labia majora. When the edges of the excised vulva are made to heal together (*infibulation*) the procedure is referred to as "Pharaonic circumcision." This obliterates access to the vaginal area except for a small opening to allow urine and menstrual blood to come out. When the woman becomes entitled to engage in coitus the orifice is enlarged by tearing it down (*introcision*).

These practices have ancient roots like male circumcision. But while male circumcision in no way incapacitates the individual, what is done in the female beyond excision of the clitoral prepuce are mutilating procedures that seriously interfere with sexual function. With all due respect to the right of cultures to fashion their own rituals, it is hard to find reasonable justification for the perpetration of these brutal practices on helpless young girls.

The Western world has had its own version of female genital mutilation. Early in the nineteenth century, "declitorization" was a medical procedure for the treatment of masturbation and "nymphomania" and was used both in Europe and the United States. Such surgery became discredited with the advent of the present century.[2]

Currently some sex therapists advocate the freeing up of clitoral adhesions in the treatment of orgasmic dysfunction. The only medical justification for clitoral excision is for conditions, such as cancer, that require radical surgery.

[1]Remy (1979).
[2]For more detailed accounts of these practices, see Hayes (1975), Huelsman (1976), and Paige (1978).

### Urethral Opening

The external urethral orifice is a small, median slit with raised margins. The female urethra conveys only urine and is totally independent of the reproductive system.

### Vaginal Opening

The vaginal orifice or *introitus* is not a gaping hole but rather is visible only when the inner lips are parted. It is easily distinguishable from the urethral opening by its larger size. The appearance of the vaginal orifice depends to a large extent on the shape and condition of the *hymen*. This delicate membrane, which only exists in the human female, has no known physiological function, but its psychological and cultural significance has been enormous (*see* Box 2.1). It varies in shape and size and may surround the vaginal orifice (annular), bridge it (septate), or serve as a sievelike cover (cribriform) (*see* Figure 2.5). There is normally always some opening to the outside.[4]

Most hymens will permit passage of a finger (or sanitary tampon), but usually cannot accommodate an erect penis without

[4]In rare instances the hymen consists of a tough fibrous tissue that has no opening (*imperforate hymen*). This condition is usually detected after a girl begins to menstruate and the products of successive menstrual periods accumulate in the vagina and uterus as an enlarging mass. It is corrected by surgical incision, usually without aftereffects.

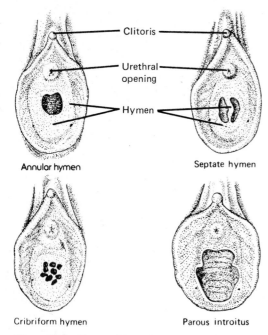

Annular hymen

Septate hymen

Cribriform hymen

Parous introitus

**Figure 2.5**   Types of hymens. From Netter, p. 90. © 1954, 1965 CIBA Pharmaceutical Company, Division of CIBA-GEIGY Corporation.

tearing. Occasionally, however, a very flexible hymen will withstand intercourse. This fact, coupled with the fact that the hymen may be torn accidentally, makes the condition of the hymen unreliable as evidence for or against virginity. In childbirth the hymen is torn further and only fragments remain attached to the vaginal opening (*see* Figure 2.5).

When the major and minor lips are removed, a ring of muscular fibers are revealed to surround the external vaginal opening (*see* Figure 2.6, in which it is shown to the left of the vagina only). Such muscular rings that act to constrict bodily orifices are known as *sphincters.* The *bulbocavernosus muscle* in this case acts as a vaginal sphincter even though it is not as highly developed as, for instance, the anal sphincter. Both voluntarily and sometimes without being aware of it, women flex this muscle and thus narrow the opening of the vagina. In normal function and in some pathological conditions, the level of

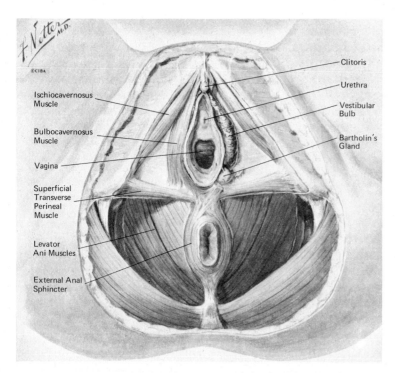

**Figure 2.6**   Female external genitalia (skin removed). From Netter, *The Ciba Collection of Medical Illustrations:* Volume 2, Reproductive System, p. 92. Reprinted by permission. © 1954, 1965 CIBA Pharmaceutical Company, Division of CIBA-GEIGY Corporation.

## Box 2.2    Defloration

The tearing of the hymen during the first coitus received a great deal of attention in "marriage manuals." Under ordinary circumstances it is an untraumatic event. In the heat of sexual excitement the woman feels minimal pain. Bleeding is generally also slight. What makes first intercourse a painful experience for some is the muscular tension that an anxious, unprepared, or unresponsive woman experiences in response to clumsy attempts at penetration. In anticipation of such difficulties some women with no premarital sexual experience used to have their hymens stretched or cut surgically before their wedding nights with the knowledge and consent of their grooms to avoid doubts of virginity or disappointment at being "cheated" of the experience of defloration. The current frequency of this practice is unknown.

The hymen is an exclusively human body part. Other primates and lower animals do not have it. Why and how the hymen evolved is not clear, but most human societies seem to have "made the most" of it. There is hardly a culture that has not been preoccupied with its proper disposal. Where defloration has been thought to pose a magical threat, special men or women have been assigned to carry it out. Among the seminomadic Yungar of Australia girls were deflowered by two old women a week before marriage. If a girl's hymen was discovered at this time to be not intact, she could be starved, tortured, mutilated, or even killed. The old custom of parading the blood-stained sheets on the wedding night as proof of the bride's chastity is well known. In various cultures, horns, stone phalluses, or other assorted implements have been used in ritual deflorations.

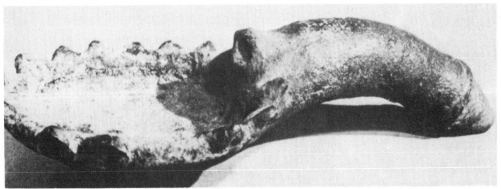

Lamp, of terracotta, used in the defloration of girls. Bastar state. Modern. From Philip Rawson, *Erotic Art of the East.* New York: G. P. Putnam's Sons, 1968. By permission.

tension exerted by these muscles is of prime importance, as we shall discuss.

Underneath the bulbocavernosus muscles are two elongated masses of erectile tissue called the *vestibular bulbs.* These structures are connected at their upper ends with the clitoris and like that organ become congested with blood during sexual arousal. They too play an important function in the female sexual response cycle and, together with the muscular ring, determine the size, tightness, and "feel" of the vaginal opening.

## Internal Sex Organs

The internal sex organs of the female consist of the paired ovaries and uterine (*fallopian*) tubes, the uterus, and the vagina, along with a few accessory structures. The ovaries have a dual function: the production of germ cells or *ova* ("eggs") and of female sex hormones (*estrogen* and *progesterone*). As hormones are secreted directly into the bloodstream, glands that produce them need no ducts and are known as *ductless,* or *endocrine,* glands.

The ovary is an almond-shaped, small ($1\frac{1}{2} \times \frac{3}{4} \times 1$ inches), and rather light organ ($\frac{1}{4}$ ounce), which shrinks further in old age. In their usual positions the ovaries lie verti-

cally (*see* Figure 2.7) flanking the uterus (*see* Figure 2.8). They are held in place by a number of folds and ligaments, including the *ovarian ligaments,* which attach them to the sides of the uterus. These ligaments are solid cords and are not to be confused with the uterine tubes, which open into the uterine cavity.

The ovary has no tubes leading directly out of it. The ova leave the organ by rupturing its wall and becoming caught in the fringed end of the fallopian tube. To permit the exit of ova, the ovarian capsule is thus quite thin. Before puberty it has a smooth, glistening surface. After the start of the ovarian cycle and the monthly exodus of ova, its

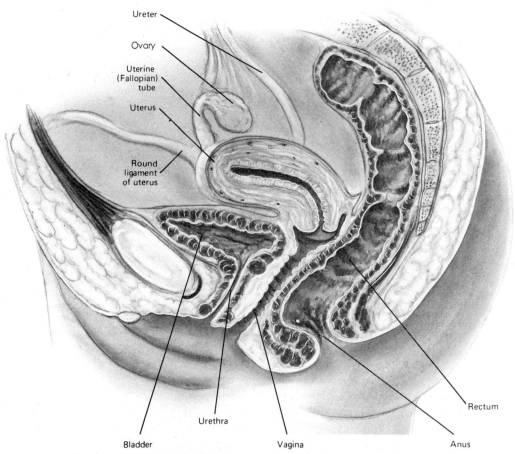

**Figure 2.7**  The female reproductive system. From Dienhart, *Basic Human Anatomy and Physiology.* Philadelphia: Saunders, 1967, p. 213. Reprinted by permission.

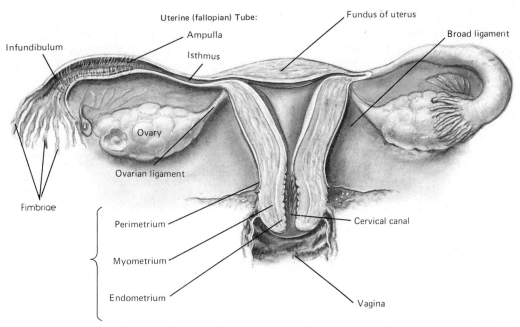

**Figure 2.8**  Internal female reproductive organs. From Dienhart, *Basic Human Anatomy and Physiology.* Philadelphia: Saunders, 1967, p. 215. Reprinted by permission.

surface becomes increasingly scarred and pitted.

The ovary contains numerous capsules, or *follicles,* in various stages of development, embedded in supporting tissues; these are located toward the periphery of the organ (the *cortex,* or "bark") (*see* Figure 2.9). The central portion of the ovary, the *medulla* ("marrow"), is rich in convoluted blood vessels.

Each follicle contains one ovum. Every female is born with about 400,000 immature ova (*oocytes*). No additional new ova are generated during the rest of a woman's life. At puberty only about 40,000 of these oocytes are left. Some of them start maturing, and each month one follicle ruptures, discharging the ovum. This process is repeated monthly and some 400 oocytes reach maturity during a woman's reproductive lifetime. The empty follicle becomes a yellowish structure (*corpus luteum*). This ovarian cycle has great reproductive and hormonal significance and will be discussed in

detail under those respective headings in Chapters 4 and 5.

The two uterine, or fallopian,[5] tubes are about 4 inches long and extend between the ovaries and the uterus. The ovarian end of the tube, the *infundibulum* ("funnel"), is cone-shaped and fringed by irregular projections, or *fimbriae,* which may cling to or embrace the ovary but are not attached to it. After leaving the ovarian surface, the ovum must find its way into the opening of the uterine tube. Although not every ovum succeeds in doing so, the process seems to be aided by a mysterious attraction between the uterine tube and the ovary. There have been instances in which women missing an ovary on one side and a uterine tube on the other have nevertheless become pregnant—all the more remarkable considering that the ovum is about the size of a needle tip and the opening

[5]Named after a sixteenth-century Italian anatomist, Gabriello Fallopio, who mistakenly thought that the tubes were "ventilators" for the uterus.

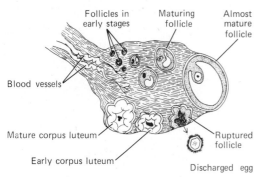

Follicles in early stages  Maturing follicle  Almost mature follicle

Blood vessels

Mature corpus luteum

Early corpus luteum

Ruptured follicle

Discharged egg

**Figure 2.9** Composite view of ovum. From Crawley, Malfetti, Stewart, and Vas Dias, *Reproduction, Sex, and Preparation for Marriage.* Englewood Cliffs, N.J.: Prentice-Hall, 1964, p. 16. Reprinted by permission.

of the uterine tube is only a slit about the size of a printed hyphen.

The second portion of the tube (the *ampulla*) accounts for over half its length. It has thin walls and is joined to the less tortuous *isthmus,* which resembles a cord and ends at the uterine border. The last segment of the tube (the uterine part) runs within the wall of the uterus itself and opens into its cavity (uterine opening).

The cavity of the uterine tube becomes progressively smaller between the ovarian

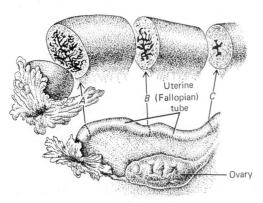

Uterine
B (Fallopian) tube

Ovary

**Figure 2.10** Uterine tube. (*A*) Infundibular, (*B*) ampullar, (*C*) isthmic. From Eastman and Hellman, *Williams Obstetrics,* 13th ed. New York: Appleton–Century–Crofts, 1966, p. 31. Reprinted by permission.

end (2 millimeters) and uterine end (1 millimeter); the numerous folds projecting into the cavity also gradually disappear (*see* Figure 2.10). The lining of the uterine tube has a deep velvety texture because of the numerous hairlike structures (*cilia*) lining it. If the ovum were the size of an orange, these cilia would be comparable in size to eyelashes.

The function of the fallopian tubes is more than mere storage and conveyance of germ cells. The fertilization of the ovum usually occurs in the infundibular third of the uterine tube, where sperm that have traversed the vagina and the uterus meet the ovum. Passage of the ovum through the tube takes several days, and, if fertilization has occurred, the structure that reaches the uterine cavity is already a complex, multicellular organism.[6]

The ovum, unlike the sperm, is not independently motile; its movement depends on the sweeping action of the cilia lining the tube and the contractions of its wall during the passage of the ovum.

Although the fallopian tubes are nowhere as surgically accessible as is the vas deferens of the male, they are still the most convenient targets for the sterilization of females. The usual procedure is to tie or sever the tubes (*tubal ligation*) on both sides. The result is sterility without concomitant impairment of sexual characteristics, desire, or ability to reach orgasm.

The uterus, or womb, is a hollow, muscular organ in which the embryo (known as the fetus after the eighth week) is housed and nourished until birth.[7] The uterus is shaped

[6]On rare occasions the fertilized ovum becomes implanted in the wall of the uterine tube itself, causing one form of ectopic ("out of normal place") pregnancy, which ultimately results in the death of the fetus and may cause the tube to rupture, with potentially serious consequences for the mother.

[7]The Greek word for uterus is *hystera,* a term that supplies the root for words like "hysterectomy" (surgical removal of the uterus) and "hysteria" (a condition believed by the ancient Greeks to result from the uterus wandering through the body in search of a child).

like an inverted pear and is usually tilted forward (anteverted) (*see* Figure 2.7).[8] The uterus is held, but not fixed, in place by various ligaments. Normally 3 inches long, 3 inches wide at the top, and 1 inch thick, it expands greatly during pregnancy. There is no other body organ that ordinarily undergoes a comparable adaptation.

The uterus consists of four parts (*see* Figure 2.8): the *fundus* ("bottom"), the rounded portion that lies above the openings of the uterine tubes; the body which is the main part; the narrow *isthmus* (not to be confused with the isthmus of the fallopian tube); and the *cervix* ("neck"), the lower portion of which projects into the vagina.

The cavity of the uterus is wider at the point at which the uterine tubes enter, but narrows toward the isthmus; the *cervical canal* then expands somewhat and narrows again at the opening (the *external os*) into the vagina (*see* Figures 2.8 and 2.11). Because the anterior and posterior uterine walls are ordinarily close together, the interior, when viewed from the side (in sagittal section), appears to be a narrow slit (*see* Figure 2.7).

The uterus has three layers (*see* Figure 2.8). The inner *mucosa,* or *endometrium,* consists of numerous glands and a rich network of blood vessels. Its structure varies with the period of life (prepubertal, reproductive, and postmenopausal) and the point in the menstrual cycle. We shall discuss these variations later in connection with sex hormones and pregnancy. The second, or muscular, layer (*myometrium*) is very well developed. Intertwined layers of smooth muscle fibers endow the uterine wall with tremendous strength and elasticity. These muscles are vital for propelling the fetus at the time of birth by means of a series of contractions. The

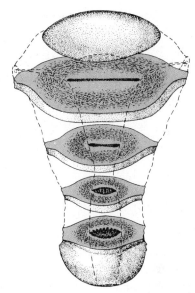

**Figure 2.11**  Reconstruction of uterus showing shape of its cavity and cervical canal. From Eastman and Hellman, *Williams Obstetrics,* 13th ed. New York: Appleton–Century–Crofts, 1966, p. 44. Reprinted by permission.

muscular layer of the uterus is continuous with the muscular sheaths of the fallopian tubes and the vagina. The isthmus of the uterus and the cervix contain fewer muscle fibers and more fibrous tissue than do the body and the fundus. The third layer, the *perimetrium* or *serosa,* is the external cover.

The vagina ("sheath") is the female organ of copulation and the recipient of the semen. Through it also pass the discharge during menstruation and the baby during birth. It does not serve for the passage of urine.

The vagina is ordinarily a collapsed muscular tube—a potential, rather than permanent, space. Its main surfaces are formed by the anterior and posterior walls, which are about 3 and 4 inches long respectively. Its side walls are quite narrow. It also appears as a narrow slit in sagittal section (*see* Figure 2.7).

The vaginal canal is slanted downward and forward. At its upper end it communi-

[8]Attempts at self-abortion or abortion by untrained individuals often end in disaster because of this anatomical feature. When a probe or long needle is introduced, the tendency is simply to push it on into the uterus; instead, the instrument pierces the roof of the vagina and penetrates the abdominal cavity.

cates with the cervical canal (usually open only the width of the lead in a pencil), and the lower end opens into the vestibule between the minor lips.

The inner lining, or vaginal *mucosa,* is like the skin covering the inside of the mouth. In contrast to the uterine endometrium it contains no glands, although its appearance is affected by hormone levels. In the adult premenopausal woman the vaginal walls are corrugated, but fleshy and soft. Following menopause they become thinner and smoother. The middle vaginal layer is muscular, but far less developed than is that of the uterine wall. Most of these fibers run longitudinally. The outer layer is also rather thin. The vaginal walls are poorly supplied with nerves so that the vagina is a rather insensitive organ except for the area surrounding the vaginal opening, which is highly excitable.

Behind the vestibular bulbs are two small glands (*Bartholin's* or *greater vestibular glands*), the ducts of which open on each side of the lower half of the vestibule in the ridges between the edge of the hymen and the minor lips. Their function is also somewhat obscure. Formerly assumed to be central in vaginal lubrication, they are now considered to play at most only a minor role in this process. The primary source of the vaginal lubricant has been shown to be the vaginal walls themselves (discussed in Chapter 3).

### Size of the Vagina

Because of its function in sexual intercourse, the vagina, like the penis, has been the subject of great interest and speculation.[9] Popular notions differentiate between tight and relaxed vaginas, those that grasp the penis and those that do not, and so on. Some aspects of these notions are demonstrable, and others are purely mythical. Functionally it is more meaningful to consider the introitus separately from the rest, for in many ways it differs from the remainder of the organ as much as the glans of the penis differs from its body.

The vagina beyond the introitus is a soft and highly distensible organ. Although it looks like a flat tube, it actually functions more as a balloon. Thus there is normally no such thing as a vagina that is permanently "too tight" or "too small." Properly stimulated, any adult vagina can, in principle, accommodate the largest penis. After all, no penis is as large as a normal infant's head, and even that passes through the vagina.

The claim that some vaginas are "too large" is more tenable. Some vaginas do not return to normal size after childbirth, and tears produced during the process weaken the vaginal walls. Even in these instances, however, the vagina expands only to the extent that the penis requires. When we add to its anatomical features the relative insensitivity of the vaginal walls, we can reasonably conclude that the main body of the vaginal cavity does not either add to or detract from the sexual pleasure of coitus in any major way. Most of the time there is no problem of "fit" between penis and vagina.

The introitus is another matter. First, it is highly sensitive. Both pain and pleasure are intensely felt there. Second, the arrangement of the erectile tissue of the bulb of the vestibule and, more important, the presence of the muscular ring of the bulbocavernosus around it make a great deal of difference in how relaxed or tight it will be. It must be emphasized that these muscles permit a significant degree of voluntary control over the size of the opening. A woman can relax or tighten the vaginal opening as she can relax or tighten the anal sphincter (though usually to a lesser extent). Furthermore, in common

---

[9]The many colloquial names for the vagina attest to the interest elicited by this organ. In *The Perfumed Garden* it has been called the "crusher," "the silent one," "yearning one," "glutton," "bottomless," "restless," "biter," "sucker," "the wasp," "the hedgehog," "the starling," "hot one," "delicious one," and so on. (Nefzawi, 1964 ed.). For more current terms see Haeberle (1978), p. 491.

with all other muscles of the body, those around the introitus can be developed by exercise.

There is continuing controversy as to whether the penis can be "trapped" inside the vagina. The prevalent view is that this is a misconception arising from the observation of dogs, in which this phenomenon occurs. The penis of the dog expands into a "knot" inside the vagina and cannot be withdrawn until ejaculation or loss of erection occurs. But, then, occasionally reports are published of the same phenomenon occurring in humans.[10]

There are other, purely psychological horrors that haunt some men: Fantasies that the vagina has teeth (*vagina dentata*) or is full of razor blades or ground glass are known and understandably influence sexual functioning.

## Breasts

The *breasts* are not part of the sex organs; however, because of their erotic significance, we shall discuss them briefly here. Breasts are characteristic of the highest class of vertebrates (mammals), which suckle their young and are therefore called *mammary glands*. Their structure and development are similar to sweat glands.

Although we generally associate breasts with females, males also have breasts that have basically the same structure but are normally not as well developed. If a male is given female hormones, he develops female-looking breasts.

The adult female breasts are located in front of the chest muscles and extend between the second and sixth ribs and from the midline of the chest to below the armpit. Each breast consists of lobes or clusters (about fifteen to twenty) of glandular tissue, each with a separate duct opening on the nipple. The lobes are separated by loosely packed fibrous

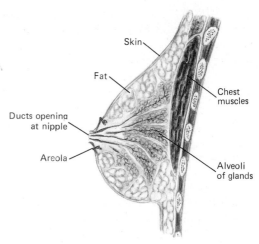

**Figure 2.12** Vertical section of the breast. From Dienhart, *Basic Human Anatomy and Physiology*. Philadelphia: Saunders, 1967, p. 217. Reprinted by permission.

and fatty tissue, which gives the breast its soft consistency (*see* Figure 2.12).

The *nipple* is the prominent tip of the breast into which the milk ducts open. It consists of smooth muscle fibers that, when contracted, make the nipple erect. The area around the nipple (*areola*) becomes darker during pregnancy and remains so thereafter. The nipple, richly endowed with nerve fibers, is highly sensitive and plays an important part in sexual arousal.

The size and shape of breasts have no bearing on their sensitivity or responsiveness. Nor is a smaller-breasted woman necessarily any less capable of breast-feeding than a larger-breasted woman. Many, but not all, women find stimulation of the breasts and nipples sexually arousing.[11]

Although the size and shape of female breasts have no physiological significance, such attributes tend to greatly influence their erotic appeal and a woman's esthetic image

---

[10]For a clinical account of a case of "penis captivus" see Melody (1977), p. 111.

[11]In addition to personal idiosyncrasy, the sensitivity of the breasts has been shown to be dependent on the hormonal levels, which fluctuate with the menstrual cycle and in pregnancy.

## Box 2.3    Breast Augmentation through Surgery

For some years plastic surgery has been used to correct breast differences and deformities that occur naturally or follow breast surgery.

The earlier techniques of breast enlargment relied on liquid silicone injections. Since these have proved unsatisfactory and have tended to lead to numerous complications, this method currently is not used by reputable surgeons. Instead, a far superior technique has been developed utilizing soft silicone implants, whereby the materials introduced into the breast are encapsulated in an inert sac and do not come into direct contact with the breast tissue. This reasonably safe approach quite successfully accomplishes the task of endowing a woman with breasts that appear and feel more satisfactory to her. There is no interference with lactation. However, such surgery is expensive, carries certain risks as do all major surgical procedures, and is objectionable to some women on ideological grounds.

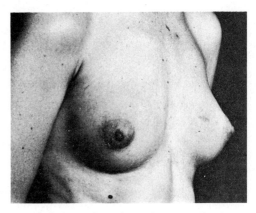

Breast augmentation through the implantation of silicone gel prosthesis. Courtesy of Dr. Donald R. Laub, Division of Reconstructive and Rehabilitation Surgery, Stanford University School of Medicine.

of herself. Such cultural judgments are of course quite arbitrary, but generally extremes in size in either direction tend to cause self-consciousness.

The female breasts develop during puberty and sometimes one grows faster than the other. The resulting asymmetry may be disturbing, but eventually the two sides become equal in size. Also, among older women, the breasts undergo other natural changes. As their supporting ligaments stretch, they tend to sag. Following menopause they become smaller and less firm. Such changes, though physiologically normal, may have psychological repercussions for some.

What to do, if anything, about these issues is a matter of personal choice. Exercises, creams, and other popularly advertised methods do not demonstrably augment breast size (apart from exercises developing the underlying chest musculature), but plastic surgery can be quite effective (*see* Box 2.3).

## Male Sex Organs

### External Genitals

The external sex organs of the male consist of the *penis* and the *scrotum* (*see* Figure 2.13).

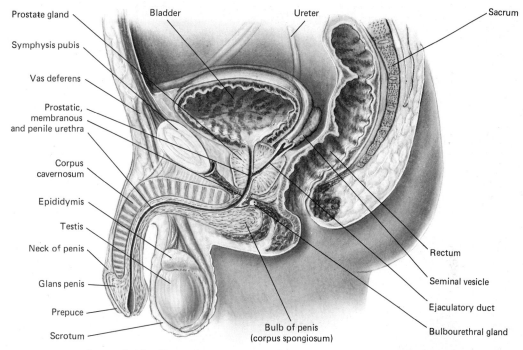

**Figure 2.13** The male reproductive system. From Dienhart, *Basic Human Anatomy and Physiology.* Philadelphia: Saunders, 1967, p. 207. Reprinted by permission.

## Penis

The penis ("tail") is the male organ for copulation and urination.[12] It contains three parallel cylinders of spongy tissue (*see* Figure 2.14) through one of which runs a tube (*urethra*) that conveys both urine and semen. The portion of the penis that is attached to the pelvis is its *root;* the free, pendulous portion of the penis is known as its *body.*

The three cylinders of the penis are structurally similar. Two of them are called the cavernous bodies (*corpora cavernosa*), and the third is called the spongy body (*corpus spongiosum*). Each cylinder is wrapped in a fibrous coat, but the cavernous bodies

have an additional common "wrapping" that makes them appear to be a single structure for most of their length. When the penis is flaccid, these bodies cannot be seen or felt as separate structures, but in erection the spongy body stands out as a distinct ridge on the underside of the penis.

As the terms "cavernous" and "spongy" suggest, the penis consists of an agglomeration of irregular cavities and spaces very much like a dense sponge. These tissues are served by a rich network of blood vessels and nerves. When the penis is flaccid, the cavities contain little blood. During sexual arousal they become engorged, and their constriction within their tough fibrous coats causes the characteristic stiffness of the penis. We shall discuss the mechanism of erection further in subsequent chapters.

At the root of the penis the inner tips (*crura*) of the cavernous bodies are attached to the pubic bones. The spongy body is not

[12]*Phallus* is the Greek name for "penis." Many colloquial terms refer to the erect penis as a pricking, probing, piercing instrument. Nefzawi's *The Perfumed Garden* has more exotic descriptions like "housebreaker," "ransacker," "rummager," "pigeon," "shamefaced one," "the indomitable," and "swimmer." (1964 ed., pp. 156–157). For more current terms see Haeberle (1978), p. 491.

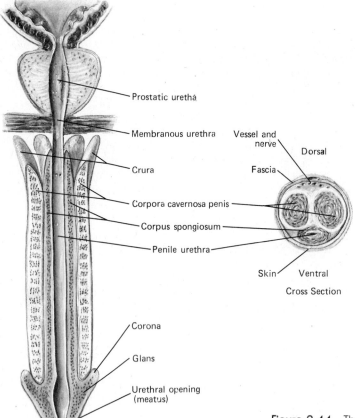

Prostatic uretha

Membranous urethra

Vessel and nerve

Dorsal

Crura

Fascia

Corpora cavernosa penis

Corpus spongiosum

Penile urethra

Skin

Ventral

Cross Section

Corona

Glans

Urethral opening (meatus)

Longitudinal Section

**Figure 2.14** The penis. From Dienhart, *Basic Human Anatomy and Physiology.* Philadelphia: Saunders, 1967, p. 211. Reprinted by permission.

attached to any bone. Its root expands to form the *bulb* of the penis, which is fixed to the fibrous "hammock" that stretches in the triangular area beneath the pubic symphysis. The crura and the bulb constitute the root of the penis.

The smooth, rounded head of the penis is known as the *glans* ("acorn") *penis.* This structure is actually formed entirely by the free end of the spongy body, which expands to shelter the tips of the cavernous bodies. The glans penis has particular sexual importance. It is richly endowed with nerves and extremely sensitive. The rest of the penis is far less sensitive. Although the glans as a whole is highly excitable, its underside, where

a thin strip of skin (*frenulum*) connects it to the adjoining body of the penis, is particularly sensitive, as is its rim, or crown (*corona*), which slightly overhangs the superficial constriction called the *neck* of the penis. This neck is the boundary between the body of the penis and the glans. At the tip of the glans is the longitudinal slit for the urethral opening (*meatus*).

The skin of the penis is hairless and unusually loose, which permits expansion during erection. Although the skin is fixed to the penis at its neck, some of it folds over and covers part of the glans, forming the *prepuce,* or *foreskin.* Ordinarily the prepuce is retractable and the glans readily exposed. Circum-

## Box 2.4   Male Circumcision

Circumcision ("cutting around") is the excision of the foreskin and is practiced around the world both as a ritual and as a medical measure. In the circumcised male the glans and the neck of the penis are completely exposed (*see* Figure 2.17).

As a ritual, circumcision was performed in Egypt as long ago as 4000 B.C. It thus long antedates the practice among Jews, Moslems, and other groups.[1] Circumcision for medical purposes in the United States dates back to the nineteenth century. Its original justification was to help combat masturbation. After this rationale was discredited, its advocates endorsed the practice on the grounds that after circumcision smegma does not accumulate under the prepuce and that it is therefore generally easier to keep the penis clean. Also, cancer of the penis was reportedly less frequent among the circumcised and cancer of the cervix less common among their spouses. Review of the evidence has shown these associations to have been highly questionable. Nevertheless,

physicians are currently still divided as to the advisability of routine circumcision in infancy.

Circumcision is medically obligatory if the foreskin is so tight that it cannot be easily retracted over the glans (*phimosis*). But this condition is rare and impossible to predict in infancy since it takes several years for the foreskin to become retractable among the majority of boys.[2]

It is generally assumed that the circumcised male is more rapidly aroused during coitus because of his fully exposed glans penis. Circumcision is also believed to cause difficulty in delaying ejaculation. Current research has failed to support these beliefs: There seems to be no difference between the excitability of the circumcised and uncircumcised penis.[3]

[1]The basis for the practice among Jews is set forth in Genesis 17:9–15. "You shall circumcise the flesh of your foreskin, and it shall be the sign of the covenant between us."
[2]For a critical review and discussion of the function of circumcision see Paige (1978).
[3]Masters and Johnson (1966), p. 190.

cision is the excision of the prepuce. In the circumcised penis, therefore, the glans is always totally exposed (*see* Box 2.4). Through more radical surgery it is also possible to form male and female genitalia through transsexual surgery (*see* Box 2.5).

Under the prepuce and in the corona and the neck are small glands that produce a cheesy substance (*smegma*) with a distinctive smell. This is a purely local secretion of no known function that must not be confused with the semen that is discharged through the urethra.

The human penis (unlike that of all other carnivora) has no bone. Nor does it have muscles within it. The *bulbocavernosus* and *ischiocavernosus* muscles surround the bulb and the crura, respectively, but their

function is primarily in helping eject urine and semen through the urethra (*see* Figure 2.15). Although these muscles may have an indirect role in contributing to venous congestion, the process of erection does not directly involve muscles, nor can the penis be moved voluntarily except for slight jerking motions.

The scrotum is a multilayered pouch. Its thin outermost skin is darker in color than is the rest of the body. It has many sweat glands, and at puberty it becomes sparsely covered with hair. The second layer consists of loosely organized muscle fibers (*dartos muscle*) and fibrous tissue. These muscle fibers are not under voluntary control, but they do contract in response to cold, sexual excitement, and a few other stimuli. Under such conditions the scrotum appears compact and

## Box 2.5  Transsexual Surgery

During the past few decades increasing attention has been attracted to individuals who for deeply felt reasons wish to undergo a radical sex change. Most often this involves males who want to become females. Much of the public interest in this connection has focused on the more dramatic physical changes brought about by surgery. Actually, where such programs are responsibly conducted, the surgiccal procedure is only one aspect of the total process of sex reassignment.

In the change from male to female the testes and most of the penile tissues are first removed. In a subsequent stage the labia and vagina are constructed to approximate as closely as possible the appearance of the female genitalia. These persons can then engage in sexual intercourse, and some of them are able to attain orgasm.

The surgical transformation from female to male is more complicated, and none of the various techniques so far devised have been as successful as the

procedures for conversion of male to female. The penis and scrotum in these cases are constructed from tissues in the genital area. The artificial penis may appear convincingly "real" in successful cases, yet it is nonfunctional in the sense of being incapable of erection. In one approach a skin tube is fashioned at the underside of the new penis where a rigid silicone tube can be inserted permitting intromission and coitus. (For a more detailed discussion of these procedures and other aspects of transsexual changes, see Laub and Gandy, 1973).

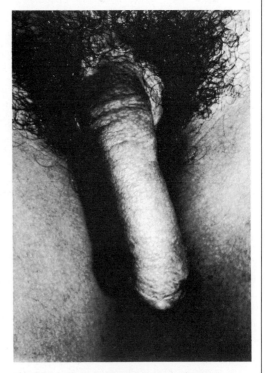

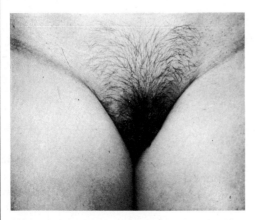

Postoperative result of male-to-female sex-change surgery. *Courtesy of Dr. Donald R. Laub, Division of Reconstructive and Rehabilitation Surgery, Stanford University School of Medicine.*

Postoperative result of female-to-male sex-change surgery. *Courtesy of Dr. Donald R. Laub, Division of Reconstructive and Rehabilitation Surgery, Stanford University School of Medicine.*

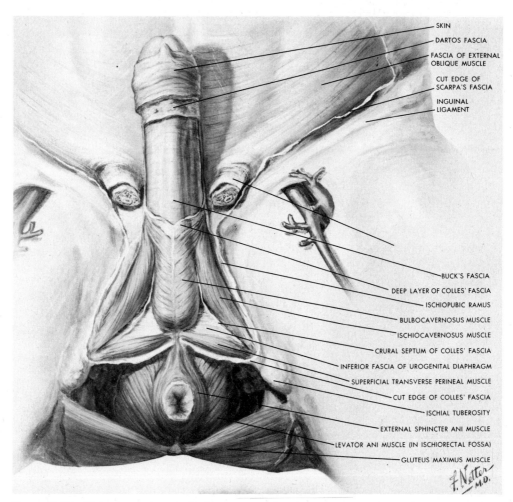

—SKIN
—DARTOS FASCIA
—FASCIA OF EXTERNAL OBLIQUE MUSCLE
—CUT EDGE OF SCARPA'S FASCIA
—INGUINAL LIGAMENT

—BUCK'S FASCIA
—DEEP LAYER OF COLLES' FASCIA
—ISCHIOPUBIC RAMUS
—BULBOCAVERNOSUS MUSCLE
—ISCHIOCAVERNOSUS MUSCLE
—CRURAL SEPTUM OF COLLES' FASCIA
—INFERIOR FASCIA OF UROGENITAL DIAPHRAGM
—SUPERFICIAL TRANSVERSE PERINEAL MUSCLE
—CUT EDGE OF COLLES' FASCIA
—ISCHIAL TUBEROSITY
—EXTERNAL SPHINCTER ANI MUSCLE
—LEVATOR ANI MUSCLE (IN ISCHIORECTAL FOSSA)
—GLUTEUS MAXIMUS MUSCLE

**Figure 2.15** Male external genitalia (skin removed). From Netter, *The Ciba Collection of Medical Illustrations:* Volume 2, Reproductive System, p. 11. Reprinted by permission. © 1954, 1965 CIBA Pharmaceutical Company, Division of CIBA-GEIGY Corporation.

heavily wrinkled. Otherwise it hangs loose and its surface is smooth. When the inner side of the thigh is stimulated, the dartos muscle contracts slightly (the *cremasteric reflex*).

The scrotal sac contains two separate compartments, each of which houses a testicle and its *spermatic cord* (*see* Figure 2.16). The spermatic cord is a composite structure from which the testicle is suspended in the scrotal sac. It includes the tube (*vas deferens*) that carries the sperm from the testicle, as well as blood vessels, nerves, and muscle fibers. When these muscles contract, the spermatic cord shortens and pulls the testicle upward within the scrotal sac.

The spermatic cord enters the abdominal cavity from the scrotal sac by traversing a region of the abdominal wall called the *inguinal canal*.

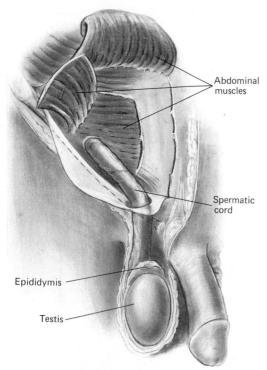

**Figure 2.16** Spermatic cord in the inguinal canal. From Dienhart, *Basic Human Anatomy and Physiology*. Philadelphia: Saunders, 1967, p. 209. Reprinted by permission.

### Size of the Penis

Variation in size and shape from individual to individual is the rule for all parts of the human body. Nevertheless, the size and shape of the penis are often the cause of curiosity and amusement as well as apprehension and concern. Representations of enormous penises can be found in numerous cultures, including some from remote antiquity (*see* Figure 2.17). These anatomical exaggerations are generally not mere caricatures or monuments to male vanity but symbols of fertility and life.

The average penis is 3 to 4 inches long when flaccid and somewhat more than 6 inches in erection. Its diameter in the relaxed state is about 1¼ inches and increases another ¼ inch in erection. Penises can, however, be considerably smaller or larger.

The size and shape of the penis, contrary to popular belief, are not related to a man's body build, race, virility, or ability to give and receive sexual satisfaction. Furthermore, variations in size tend to decrease in erection: The smaller the flaccid penis, the proportionately larger it tends to become when erect. The penis does not grow larger through frequent use.

Symbolic representations of the penis have been used for religious and magical functions (*see* Boxes 2.6 and 2.7).

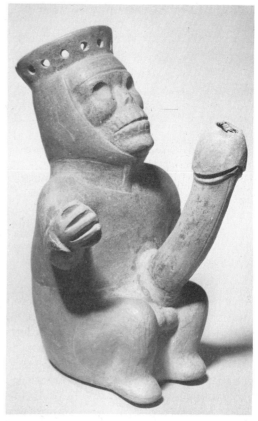

**Figure 2.17** Mochica pottery. Courtesy of William Dellenback, Institute for Sex Research, Inc.

**Box 2.6**    Phallus Worship

The worship of the male genitals is one of the oldest religious practices known. It is usually interpreted as related to fertility cults: the expression of man's desire for the perpetuation of the race and his identification with the reproductive powers of nature. In ancient Greece phallic worship centered around Priapus and the Dionysiac cults. Priapus was the son of Aphrodite (goddess of love) and Dionysus (god of fertility and wine). Usually represented as a grinning little man with an enormous penis, Priapus was very much in evidence during the many festivals honoring Dionysus, which were occasions for orgiastic abandon.

Under the Roman Empire phallic worship took on a less festive, rather grim form. During the yearly festival the notorious Day of Blood, some frenzied participants actually mutilated their own genitals. In fact, self-castration during this festival became a prerequisite for admission into the priesthood.

In India, Shiva (or Siva), one of the three supreme gods of Hinduism, was symbolically represented and worshipped as an erect penis (the lingam).

The phallic symbol could stand above or combined with a symbol of the female vulva (the yoni) as shown in the accompanying figure. Offerings of milk would be poured on the lingam, then drained off from the surrounding basin and used for various ritual purposes. Phallus worship has also been prominent in Japanese and some American Indian fertility rites.

Lingam-yoni altar. Stone. Eastern India. Modern. From Philip Rawson, *Erotic Art of the East.* New York: G. P. Putnam's Sons, 1968.

## Internal Sex Organs

The internal sex organs of the male consist of a pair of testes or testicles, with their duct system for the storage and transport of sperm (epididymis, vas deferens, and ejaculatory duct in pairs and the single urethra) and accessory organs (a pair of seminal vesicles and bulbourethral glands and the prostate gland).

The testes ("witnesses": from the ancient custom of placing the hand on the genitals when taking an oath) are the *gonads*, or reproductive glands, of the male. They produce sperm, as well as the male hormone testosterone.

The two testicles are about the same size ($2 \times 1 \times 1\frac{1}{4}$ inches), although the left one usually hangs somewhat lower than the right one. The weight of the testicles varies from one person to another, but averages about 1 ounce and tends to lessen in old age.

## Box 2.7  Magical Uses of Genital Symbols

Since paleolithic times, various symbolic representations of the male and female genitalia have been put to magical uses. The Romans used phallic symbols in numerous guises to ward off the "evil eye." Sometimes these consisted of small stone phalluses embedded in a wall or larger free-standing representations placed at the boundary of the property. Commonly, they took the form of amulets showing the penis in naturalistic or fantastic forms. The magical thinking behind the practice was that such objects would fascinate the person with the evil eye and divert his gaze from the wearer of the amulet, thus sparing the person.[1] Similar practices in more or less obvious form can be discerned in other past and present cultures.

[1]For further details, see Grant (1975).

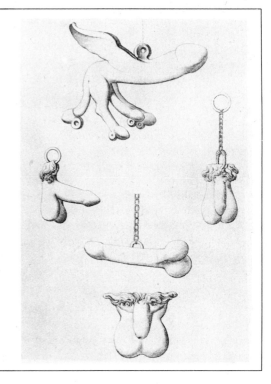

Each testicle is enclosed in a tight, whitish fibrous sheath (*tunica albuginea,* "white tunic")[14] which at the back of the organ thickens (*mediastinum testis*) and penetrates the testicle. Its ramifications (*septa*) then spread out within the organ and subdivide it into conical *lobes* (*see* Figure 2.18). Each lobe is packed with convoluted *seminiferous* (sperm-bearing) *tubules.* These threadlike structures are the sites at which sperm are produced. Each seminiferous tubule is 1–3 feet long, and the combined length of the tubules of both *testes* measures several hundred yards.

[14]This anatomical feature is responsible for the sterility that may follow mumps in adulthood. When this virus infection involves the testicles, the swelling organs push against their unyielding covers, and the pressure destroys the delicate tubes in which the sperm develop. When the same virus infects the female ovaries, the organs simply swell up and then return to their normal size and function.

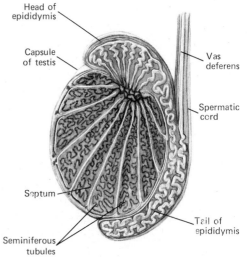

**Figure 2.18**  Testis and epididymis. From Dienhart, *Basic Human Anatomy and Physiology.* Philadelphia: Saunders, 1967, p. 207. Reprinted by permission.

This elaborate system of tubules allows for the production and storage of hundreds of millions of sperm. The process of *spermatogenesis,* or sperm production (discussed more fully in Chapter 5), takes place exclusively within the seminiferous tubules: Microscopic cross sections thus shown sperm at various levels of maturation. The first or earliest cell in this maturational chain is the *spermatogonium,* which in subsequent stages of development is successively called a *spermatocyte* (primary and secondary); *spermatid;* and fi-

nally, when mature, *spermatozoan,* or sperm. The seminiferous tubules of the newborn are solid cords in which only undifferentiated cells are evident. Following puberty, the tubules develop a hollow center into which the sperm are released (*see* Figure 2.19).

The second major function of the testes is the production of the male sex hormone. The testicular cells that produce the male hormone are found between the seminiferous tubules and are therefore known as *interstitial cells* (or *Leydig's cells*): They are scattered in

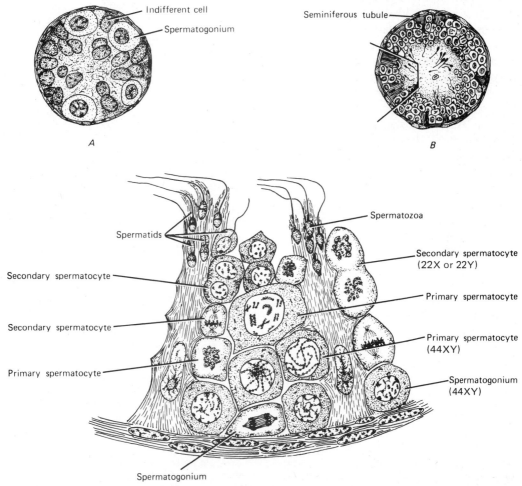

**Figure 2.19** Human testis tubules, in transverse section. (*A*) Newborn (×400); (*B*) adult (×115); (*C*) detail of the area outlined in (*B*) (×900). From Arey, *Developmental Anatomy,* 7th ed. Philadelphia: Saunders, 1965, p. 41. Reprinted by permission.

the connective tissue in close association with blood vessels. The cells responsible for the two primary functions of the testes (reproductive and endocrine) are thus quite separate and never in contact.

The seminiferous tubules converge in an intricate maze of ducts, which ultimately leave the testis and fuse into a single tube, the epididymis. Figure 2.20 shows the route that sperm follow through these larger ducts during ejaculation.

The epididymis ("over the testis") constitutes the first portion of this paired duct system. Each is a remarkably long tube (about 20 feet) that is, however, so tortuous and convoluted that it appears as a C-shaped structure not much longer than the testis to whose surface it adheres (*see* Figure 2.18).

The vas deferens, or *ductus deferens* ("the vessel that brings down"), is the less tortuous and much shorter continuation of the epididymis. It travels upward in the scrotal sac for a short distance before entering the

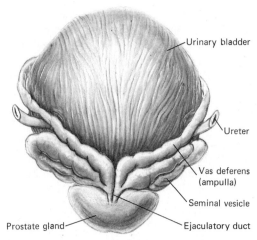

**Figure 2.21**   The bladder, seminal vesicles, and prostate gland—posterior view. From Dienhart, *Basic Human Anatomy and Physiology.* Philadelphia: Saunders, 1967, p. 210. Reprinted by permission.

abdominal cavity; its portion in the scrotal sac can be felt as a firm cord.[15]

The terminal portion of the vas deferens is enlarged and once again tortuous; it is called the *ampulla* ("flask"). It passes to the back of the urinary bladder (*see* Figure 2.21), narrows to a tip, and joins the duct of the seminal vesicle to form the *ejaculatory duct.* This portion of the paired genital duct system is very short (less than 1 inch) and quite straight. It runs its entire course within the *prostate gland* and opens into the prostatic portion of the *urethra* (*see* Figures 2.13 and 2.21).

The *urethra* has a dual function in the

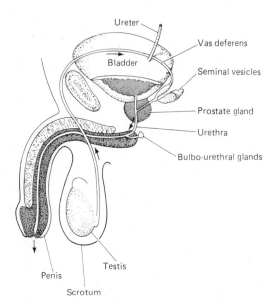

**Figure 2.20**   Passage of spermatozoa. From McNaught and Callander, *Illustrated Physiology.* Baltimore: Williams & Wilkins, 1972, p. 178. Reprinted by permission.

[15]That the vas is so easily located and surgically accessible makes it the most convenient target for sterilizing men. The operation (vasectomy, "cutting the vas") is very simple: under local anesthesia, a small incision is made in the scrotum, and the vas is simply either tied or cut. To stop all passage of sperm, the operation must, of course, be performed on both sides. The man is then sterile, but his sex drive, potency, and ability to have orgasms are not affected. He will continue to ejaculate, but his semen will contain no sperm. It is sometimes possible to reestablish fertility by untying or suturing the severed vas. This procedure is, however, very delicate and not always successful.

male, conveying both semen and urine. It begins at the bottom of the urinary bladder and periodically empties accumulated urine. (It must not be confused with the two *ureters,* each of which begins in one kidney and carries urine into the bladder.) The urethra is about 8 inches long and is subdivided into prostatic, membranous, and penile parts (*see* Figures 2.13 and 2.14).

The prostatic portion of the urethra is more easily dilated than are the others. In its posterior wall are the tiny openings of the two ejaculatory ducts. The multiple ducts of the prostate gland also empty into the prostatic urethra like a sieve (*see* Figure 2.14).

Voluntary control of urination is made possible by the muscle fibers (*urethral sphincter*) surrounding the short membranous urethra. When sufficient urine has accumulated in the bladder, the resulting discomfort prompts the person to relax the urethral sphincter to allow the passage of urine. The external urethral opening has no sphincter, which is why at the end of urination the urine left in the penile portion must be squirted out through the contraction of the bulbocavernosus and ischiocavernosus muscles.

The penile part of the urethra has already been discussed. It pierces the bulb of the corpus spongiosum, traverses its whole length, and terminates at the tip of the glans in the external urethral opening (*see* Figure 2.14). The *bulbourethral glands* (to be discussed later) empty into this portion.

## Accessory Organs

Three accessory organs perform auxiliary functions in the male. They are the prostate gland, two seminal vesicles, and two bulbourethral glands.

The prostate is an encapsulated structure about the size and shape of a large chestnut and consisting of three lobes. It is located with its base against the bottom of the bladder. It consists of smooth muscle fibers and glandular tissue whose secretions account for

much of the seminal fluid and its characteristic odor. As we have described, it is traversed by the urethra and the two ejaculatory ducts, and prostatic fluid is conveyed into this system through a "sieve" of multiple ducts.

The prostate is small at birth, enlarges rapidly at puberty, but usually shrinks in old age. Sometimes, however, it becomes enlarged and interferes with urination, necessitating surgical intervention. (It may be removed piece by piece through the urethra or in open surgery.) The size of the prostate is determined clinically by means of rectal examination.

The seminal vesicles are two sacs, each about 2 inches long. Each ends in a straight, narrow duct, which joins the tip of the vas deferens to form the ejaculatory duct. The function of the seminal vesicles was once assumed to be the storage of sperm (each holds about 2 to 3 cubic centimeters of fluid), but it is currently believed to be primarily involved in contributing fluids that initiate the motility of sperm through the action of their tails.

The bulbourethral glands (*Cowper's glands*) are two pea-sized structures flanking the penile urethra into which each empties through a tiny duct. During sexual arousal these glands secrete a clear, sticky fluid that appears as a droplet at the tip of the penis (the "distillate of love"). There usually is not enough of this secretion to serve as a coital lubricant. As it is alkaline, however, it may help to neutralize the acidic urethra, which is harmful to sperm, in preparation for the passage of semen. Although this fluid must not be confused with semen, it does often contain stray sperm, which explains pregnancies resulting from intercourse without ejaculation.

## Developmental Anatomy of the Sex Organs

The study of the developmental anatomy of the various systems and organs of the body is a separate science (embryology). A com-

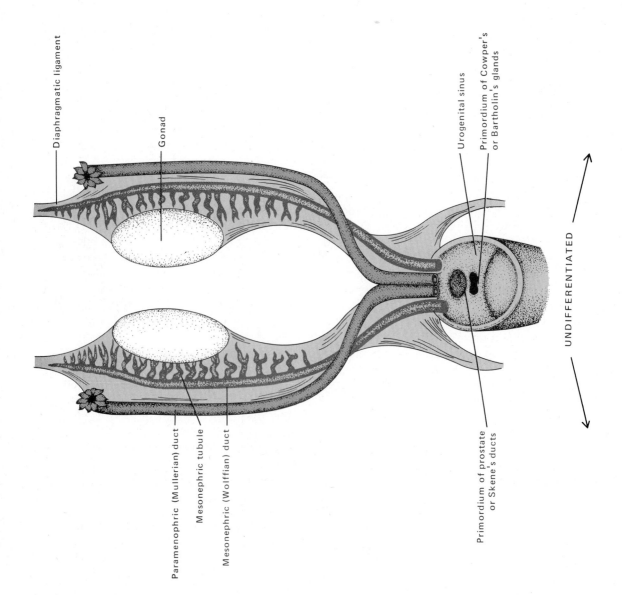

Diaphragmatic ligament

Gonad

Urogenital sinus

Primordium of Cowper's
or Bartholin's glands

Paramenophric (Mullerian) duct

Mesonephric tubule

Mesonephric (Wolffian) duct

Primordium of prostate
or Skene's ducts

UNDIFFERENTIATED

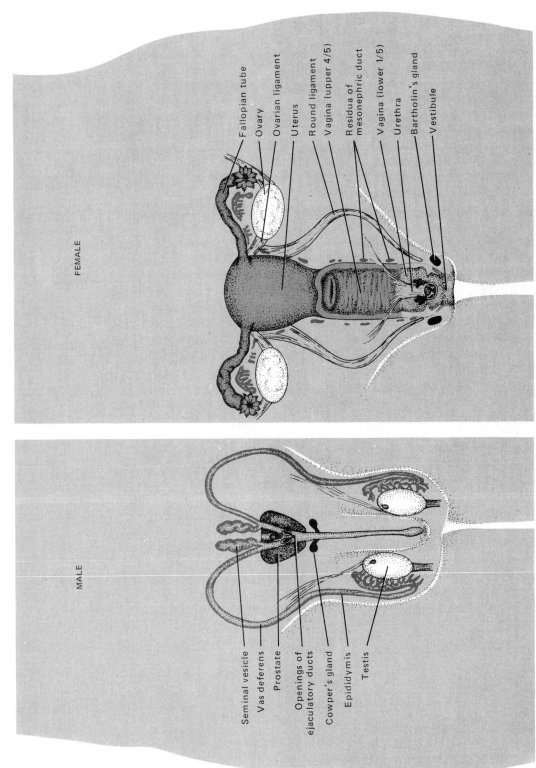

MALE

Seminal vesicle
Vas deferens
Prostate
Openings of
ejaculatory ducts
Cowper's gland
Epididymis
Testis

FEMALE

Fallopian tube
Ovary
Ovarian ligament
Uterus
Round ligament
Vagina (upper 4/5)
Residua of
mesonephric duct
Vagina (lower 1/5)
Urethra
Bartholin's gland
Vestibule

**Figure 2.22**   Homologues of internal male and female genitalia; development from undifferentiated into differentiated stage. From Netter, *Reproductive System*, The Ciba Collection of Medical Illustrations, vol. 2 Summit, N.J.: Ciba, 1965, p. 2. Reprinted by permission. © 1954, 1965 CIBA Pharmaceutical Company, Division of CIBA-GEIGY Corporation.

prehensive survey of the development of the reproductive system is far too complex an undertaking for us to attempt here. We shall therefore select only a few aspects of this process to illustrate how sex organs develop and to emphasize the basic *structural* similarities between the male and female reproductive systems. Intrauterine growth and development in general will be discussed in conjunction with conception and pregnancy.

The genital system makes its appearance during the fifth to sixth week of intrauterine life, when the embryo has attained a length of 5 to 12 millimeters. At this undifferentiated stage the embryo has a pair of gonads and two sets of ducts (*see* Figure 2.22), as well as the rudiments of external genitals (*see* Figure 2.23).

At this time one cannot reliably determine the sex of the embryo by either gross or microscopic examination. The gonads have not yet become either testes or ovaries, and the other structures are also undifferentiated. This lack of visible differentiation does not mean that the sex of the individual is still undecided; sex is determined at the very moment of fertilization and depends on the chromosome composition of the fertilizing sperm. If the fertilizing sperm carries a Y chromosome, the issue will be male. Otherwise it will be female. These mechanisms that determine the process of sex differentiation will be discussed in Chapters 4 and 5.

## Differentiation of the Gonads

The gonad that is destined to develop into a testis gradually consolidates as a more compact organ under the influence of the Y chromosome. Some of its cells are organized into distinct strands (*testis cords*), the forerunners of the seminiferous tubules. Other structures form the basis of the future internal duct system. By the seventh week (when the embryo is 14 millimeters long) enough differentiation may have occurred so that the organ is recognizable as a developing testis. If by this time

the basic architecture of the future testis is not discernible, it can be provisionally assumed that the undifferentiated gonad will develop into an ovary. More definitive evidence that the baby will be a girl comes somewhat later (about the tenth week), when the forerunners of the follicles begin to become visibly organized. After these basic patterns are set, the testis and ovary continue to grow accordingly. These organs do not attain full maturity until after puberty. The differences between the gonads of the newborn and the adult of each sex, briefly outlined above, will be further elaborated in Chapters 4 and 5.

## Descent of the Testis and Ovary

Concomitant with the development described, both the testis and the ovary undergo gross changes in shape and position that are of special significance. At first the testis and the ovary are slender structures high up in the abdominal cavity. By the tenth week they have grown and shifted down to the level of the upper edge of the pelvis. There the ovaries remain until birth; they subsequently rotate and move farther down until they reach their adult positions in the pelvis.

In the male this early internal migration is followed by the actual descent of the testes into the scrotal sac (*see* Figure 2.24). After the descent of the testes, the passage is obliterated.

Two clinical problems may arise during this process. First, one or both of the testes may fail to descend into the scrotum before birth, as happens in about 2 percent of males born. In most of these boys the testes do descend by puberty. However, if they do not do so spontaneously, hormonal or surgical intervention becomes necessary. Otherwise the higher temperature of the abdominal cavity would interfere with spermatogenesis, resulting in sterility if both testes have failed to descend into the scrotum. Undescended testes are also more likely to develop cancer.

The second problem arises when the

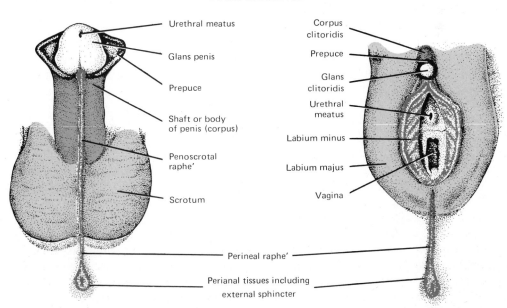

**Figure 2.23** Homologues of external male and female genitalia; development from undifferentiated into differentiated stage. From Netter, *Reproductive System*, The Ciba Collection of Medical Illustrations, vol. 2 Summit, N.J.: Ciba, 1965, p. 3. Reprinted by permission. © 1954, 1965 CIBA Pharmaceutical Company, Division of CIBA-GEIGY Corporation.

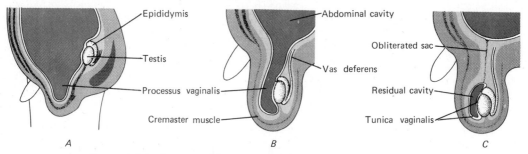

Epididymis

Testis

Processus vaginalis

Cremaster muscle

Abdominal cavity

Vas deferens

Obliterated sac

Residual cavity

Tunica vaginalis

A                                    B                                    C

**Figure 2.24**   Descent of human testis and its subsequent relations shown in diagrammatic hemisections. From Arey, *Developmental Anatomy,* 7th ed. Philadelphia: Saunders, 1965, p. 333. Reprinted by permission.

passage traversed by the testes is not eliminated or reopens when the tissues become slack in old age. An abnormal passage is created, and intestinal loops may find their way into the scrotal sac creating a condition known as *inguinal hernia,* or *rupture* (*see* Figure 2.25). It too can readily be corrected by surgery.

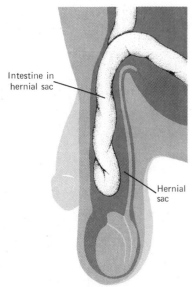

Intestine in hernial sac

Hernial sac

**Figure 2.25**   Inguinal hernia. From Arey, *Developmental Anatomy,* 7th ed. Philadelphia: Saunders, 1965, p. 334. Reprinted by permission.

## Differentiation of the Genital Ducts

In the undifferentiated stage the embryo has two sets of ducts: the *paramesonephric* or *Müllerian* (the potential female) and the *mesonephric,* or *Wolffian* (the potential male) ducts (*see* Figure 2.24).

Just as the Y chromosome directs the development of the undifferentiated gonad to become a testis, the embryonal testis in turn determines the future development of the genital ducts. It accomplishes this by producing two substances: fetal androgen, which promotes the further differentiation of the Wolffian system, and a second substance that inhibits the further development of the Müllerian ducts. As a result the Wolffian duct on each side eventually becomes the epididymus, vas deferens, and seminal vesicle, and the Müllerian ducts degenerate.

In the absence of these testicular hormones (and without the need of ovarian hormones) the Wolffian ducts degenerate and the Müllerian ducts form the fallopian tubes, uterus, and the upper two-thirds of the vagina.

The lower third of the vagina, the bulbourethral glands, and the urethra in both sexes and the prostate gland are derived from the urogenital sinus which is part of the urinary system.

The precise mechanisms involved in the

above process are not entirely clear, but the general principle holds that differentiation of the reproductive tract is dependent on the Y chromosome first and then on testicular hormones. Otherwise the undifferentiated system will develop into the female pattern. Thus, irrespective of genetic makeup, if the gonads are removed early in life, the organism will develop into a female.

## Differentiation of the External Genitals

The external genitals, like the internal, are at first sexually undifferentiated (*see* Figure 2.23). As we have indicated, by the seventh week the sex of the embryo can be ascertained provisionally by means of a microscopic study of the gonads. Several more weeks are necessary for the more distinctive development of the external genitals. The sex of the 4-month-old fetus is unmistakable. Differentiation of male and female external genitals is illustrated in Figure 2.23.

As with the differentiation of the internal duct system, it is testicular hormones that convert the undifferentiated precursors of the external genitalia to the male pattern. Otherwise it is the female pattern that develops.

This process of differentiation usually proceeds in a predictable and orderly manner. However, anomalies of development occasionally result in paradoxical conditions of bisexuality or hermaphroditism (discussed in Chapter 4).

With the aid of Figures 2.22 and 2.23 it is now easier to match the homologous pairs of organs in male and female. In principle it should be possible to identify the homologue to every part; however, since the degenerated remnants of theWolffian ducts in the female and the Müllerian ducts in the male are inconsequential structures, there is no point in attempting this for our purposes. The homologous pairs of significance are as follows: seminiferous tubules–ovarian follicles; Bartholin's gland–Cowper's gland; glans penis–glans of clitoris; corpora cavernosa of penis–corpora cavernosa of clitoris; corpus spongiosum of penis–bulb of the vestibule; underside of penis–labia minora; scrotum–labia majora.[16]

[16]A complete listing of all homologous pairs may be found in Moore (1973), p. 216.

# Chapter 2

# Physiology of Sexual Functions

The physiological goal of sexual activity is the attainment of orgasm. A great deal of such activity falls short of this goal, and the frequency and means by which orgasm is achieved through a lifetime vary tremendously. In studies of sexual behavior, orgasm is used as a quantitative unit, for it can be easily recognized and its occurrence counted.

In this chapter we are primarily concerned with physiological processes. There is an inherent danger in this approach, and it can be legitimately asked whether or not it is ever possible to separate the physiological functions of the body from their psychological concomitants. We do not intend to claim a dichotomy between mind and body, but practical considerations dictate that we examine the various facets of human sexuality, a few at a time. Furthermore, as Kinsey has stated: "Whatever the poetry and romance of sex and whatever the moral and social significance of human sexual behavior, sexual responses involve real and material changes in the physiologic functioning of an animal."[1]

It is common knowledge that an epi-

sode of sexual experience begins with mounting excitement, which culminates in orgasm and is followed by a period of relaxation and sexual satiety. In more formal descriptions various stages have been ascribed in this process. Havelock Ellis envisaged two phases: tumescence and detumescence. More recently Masters and Johnson have proposed four phases: excitement, plateau, orgasm, resolution. The purpose of such subdivisions is to facilitate observation and description. Such classifications are not intended to obscure the basic unity of sexual activity, which proceeds in a continuous manner with manifestations of various phases overlapping and merging into one another imperceptibly.

## Sexual Stimulation

The ability to respond to sexual stimulation is a universal characteristic of all healthy beings. Although the nature of the stimulus varies widely, the basic physiological response of the body is the same. The varieties and intensity of sexual arousal that each person experiences during a lifetime are, however, legion. Sometimes this excitement achieves full expression in orgasm; most of

[1]Kinsey et al. (1953), p. 594.

the time it progresses no farther than lingering thoughts or vague, evanescent yearnings.

What triggers such responses? Potentially, anything and everything. The stimuli may be "sexual" in the ordinary sense of the term; or they may involve factors that have no erotic interest for most people. The number and variety of sexual stimulants are bewildering. To impose some order, it is usual to classify erotic stimuli on different parameters: physical versus psychological, those to which responses are innate versus those to which they are learned, and so on. Such classification attempts are, however, of limited usefulness, for most stimuli have both physical and psychic components, and all behavior has internal and external determinants.

All modalities of sensation can be and usually are involved in erotic arousal, but for most human beings touch predominates, followed by vision. Among other animals other sensory modalities, like smell and taste (as in insects), may be more potent.

## Stimulation through Touch

Tactile stimulation probably figures in most instances of sexual arousal leading to orgasm. Though other modalities are also important, touch remains the predominant mode of erotic stimulation. It is, in fact, the only type of stimulation to which the body can respond reflexively and independently of higher psychic centers. Even a man who is unconscious or whose spinal cord is injured at a point that prevents impulses from reaching the brain (but leaves sexual coordinating centers in the lower spinal cord intact) can still have an erection when his genitals or inner thighs are caressed.

The perception of touch is mediated through nerve endings in the skin and deeper tissues. These end organs are distributed unevenly, which explains why some parts (fingertips) are more sensitive than others (the skin of the back): The more richly innervated the region, the greater is its potential for stimulation.

Some of the more sensitive areas are believed to be especially susceptible to sexual arousal and are therefore called *erogenous zones.* They include the clitoris; the glans penis (particularly the corona and the underside of the glans); the shaft of the penis; the labia minora and the space they enclose (vestibule); the vaginal canal; the area between the anus and the genitals; the anus itself; the breasts (particularly the nipples); the mouth (lips, tongue, and the whole interior); the ears (especially the lobes); the buttocks; and the inner surfaces of the thighs.

Based on questionnaire responses, the primary nongenital erotic zones are reportedly the thighs (32 percent males, 14 percent females), the lips (10 percent males, 6 percent females) and the breasts (36 percent females, 5 percent males).[2]

Although it is true that these areas are most often involved in sexual stimulation, they are not the only ones by any means. The neck (throat and nape), the palms and fingertips, the soles and toes, the abdomen, the groin, the center of the lower back, or any other part of the body may well be erotically sensitive to touch. Some women have reached orgasm when their eyebrows were stroked or when pressure was applied to their teeth alone.[3]

The concept of erogenous zones is not new: Explicit or implicit references to them are plentiful in "love" manuals, and the practical value of such knowledge is self-evident.[4] A knowledge of erogenous zones may greatly enhance one's effectiveness as a lover. It must be noted, however, that these zones are often indistinct and do not correspond to any given pattern of nerve distribution. Also, the ultimate interpretation of all stimuli by the brain is profoundly affected by previous ex-

[2]Goldstein (1976), p. 130
[3]Kinsey et al. (1953), p. 590.
[4]The *Kama Sutra* refers to the armpit, throat, breasts, lips, "middle parts of the body," and thighs as suitable locations for stimulation (Vatsyayana, 1963 ed., p. 99). Also see Box 3.1.

## Box 3.1 "Of the Various Seats of Passion in Women"

In classical Hindu erotic manuals, such as the *Ananga Ranga*, detailed instructions are given on the erogenous female zones and how and when to stimulate them. In addition to the general table reproduced below, more specific tables give instruction for the arousal of women of different classes of temperament. Although such prescriptions have no demonstrable physiological validity, they show the concerted effort to bring together various beliefs on human nature and cosmic forces in the service of sexuality.

From Richard Burton and F. F. Arbuthnot, Trans., *The Ananga Ranga or the Hindu Art of Love of Kalyana Malla*. New York: Putnam, 1964.

| Shuklapaksha or light fortnight; right side | | The touches by which passion is satisfied | Krishnapaksha or dark fortnight; left side | |
|---|---|---|---|---|
| Day | Place | | Place | Day |
| 15th | Head and hair | Hold hair, and caress the head and fingertips | Head and hair | 1st |
| 14th | Right eye | Kiss and fongle | Left eye | 2nd |
| 13th | Lower lip | Kiss, bite and chew softly | Upper lip | 3rd |
| 12th | Right cheek | Do. | Left cheek | 4th |
| 11th | Throat | Scratch gently with nails | Throat | 5th |
| 10th | Side | Do. | Side | 6th |
| 9th | Breasts | Hold in hands and gently knead | Breasts | 7th |
| 8th | All bosom | Tap softly with base of fist | All bosom | 8th |
| 7th | Navel | Pat softly with open palm | Navel | 9th |
| 6th | Nates | Hold, squeeze and tap with fist | Nates | 10th |
| 5th | Yoni | Work with friction of Linga | Yoni | 11th |
| 4th | Knee | Press with application of knee and fillip with finger | Knee | 12th |
| 3rd | Calf of leg | Press with application of calf and fillip with finger | Calf of leg | 13th |
| 2nd | Foot | Press with toe, and thrust the latter | Foot | 14th |
| 1st | Big toe | Do. | Big toe | 15th |

perience and conditioning. A specific "erogenous" zone may thus be quite insensitive in a given person, or it may be sensitive to the point of pain. One cannot therefore approach another person in a mechanical, push-button manner and expect to elicit automatic sexual arousal.

Although it is true that one is more likely than not to respond to stimulation of the usual erogenous zones, the subtle lover will seek to learn the unique erogenous map of his or her partner, which is the result of both biological endowment and life experiences.

The frankly erotic component in being touched must be understood in the broader, more fundamental need for bodily contact that stems from our infancy. A crucial component in infant care is the touching, caressing, fondling and cuddling that is carried out by the adult caretakers. The dire effects that result from the deprivations of such care have been well documented for humans[5] and other primates.[6] Indeed, tactile communication plays such a major part in primate life

[5]Spitz and Wolf (1947), Harrow and Harmon (1975).

[6]Harlow (1958). We shall discuss some of Harlow's work in Chapter 8.

that the practice of grooming has been called the "social cement of primates."[7] The use of touch is a good example of how an ostensibly sexual process, in this case arousal, is embedded in the more fundamental human needs for contact, security, and affection.

## Stimulation through Other Senses

Vision, hearing, smell, and taste are to varying extents also important avenues of erotic stimulation. These modalities, in contrast to touch, do not operate reflexively. We learn to experience certain sights, sounds, and smells as erotic and others as neutral or even offensive. We are not born with the notion that roses have an attractive odor or that feces (particularly those of others) smell bad. Regardless of the claims of cosmetic manufacturers, there are no scents and colors that are "naturally erotic," but scents and colors can be exciting if we have been conditioned to associate them with sexual arousal.[8]

The reflexive basis of tactile stimulation does not prevent its being subject to experiential modification. As discussed earlier, any part of the body surface may be rendered erotically sensitive or insensitive through experience and mental associations. Yet usually the reflexive aspect of tactile stimulation continues to operate. With vision, hearing, and smell, all responses are learned.

There is consequently boundless diversity in the sexual preferences and dislikes of individuals, as well as of cultures. This diversity makes it impossible to generalize about the effectiveness of any one source of stimulation. Why a certain female profile or a male sexual characteristic should be found exciting in one culture but not in another or only during a certain period of a given culture is cause for endless speculation.

The sight of the female genitals is prob-ably as nearly a universal source of excitement for heterosexual males as any that exists. Paradoxically, viewing the male genitals does not seem to excite females as much, but this may simply be due to cultural inhibitions. There certainly are women who are just as fascinated by male genitals as men are by female genitals. In cultures in which nudity is acceptable, open preferences for certain features of external sex organs develop. Among some peoples of South Africa, for instance, large, pendulous minor lips ("Hottentot aprons") were considered very attractive in women. By pulling and stretching these parts during childhood and adolescence women caused them to hypertrophy. Alteration and mutilation of the genitals do not, however, always have erotic purposes, but may serve magical or religious functions.

Even though erotic standards differ and change, the impact of visual stimuli are unmistakable, as our preoccupation with physical attributes, cosmetics, and dress indicates. We could argue whether or not esthetic concerns are erotically motivated in all instances. Furthermore, the ability to experience and elicit sexual feelings does not necessarily depend on shape and form, as experience with beautiful but unresponsive women or handsome but inept men readily demonstrates.

The effect of sound is perhaps less telling but nevertheless quite significant as a sexual stimulant. The tone and softness of voice as well as certain types of music (with pulsating rhythms or repeated languorous sequences) can serve as erotic stimuli. But since these responses are learned, what stimulates one person may simply distract or annoy another.

The importance of the sense of smell has declined in man, both generally and in sexual terms. Nevertheless, the use of scents in many cultures, as well as preoccupation with body odors, attests to its considerable influence.[9]

[7]Jolly (1972), p. 153).

[8]An intriguing possibility exists that humans, like some other animals, secrete pheromones, in which case the above statement will have to be modified. Pheromones are discussed in Chapter 4.

[9]For further discussion of the role of olfaction, see Schneider (1971).

We would expect the smell of vaginal secretions and semen to have erotic properties. Yet most people do not openly admit to recognizing such properties—possibly because of inhibitions about discussing sexual matters. Observation of sexual behavior among animals indicates not only that our sense of smell has declined but also that we have lost our enjoyment of whatever body smells we do perceive. Sexual stimulants will be discussed further in connection with aphrodisiacs in Chapter 4.

## Emotional Stimulation

Despite all the fascinating information that we have about the physical basis of sexual stimulation, the key to understanding human sexual arousal remains locked in emotional processes. Sexual arousal is an emotional state that is greatly influenced by other emotional states. Even reflexive sexual elements are normally under emotional control. Stimulation through any or all of the senses will result in sexual arousal if, and only if, accompanied by the appropriate emotional concomitants. Certain feelings like affection and trust will enhance, and others like anxiety and fear will inhibit erotic responses in most cases.

Given our highly developed nervous systems, we can react sexually to purely mental images, which makes sexual fantasy the most common erotic stimulant. Our responsiveness is thus not solely based on the characteristics of the concrete situation, but includes the entire store of memory from past experience and thoughts projected into the future. What arouses us sexually is in the last analysis the outcome of all these influences.

As human beings we all share to some extent a common history, as well as a common biology. For example, we are all cared for as infants by older persons who become the first and most significant influences in our lives. But we are also unique in numerous ways. The earlier remarks about erogenous zones can therefore be applicable to the psychological realm as well as to the physical.

Just as most of us are likely to respond to gentle caressing on the inside of our thighs, we are also likely to respond positively to the expression of sexual interest in us. In both cases the response will depend on the person involved and on the circumstances, and is hardly an automatic reaction.

Much of what is being said here is rather obvious, but it needs to be reiterated because we often find it easier and less threatening to rely on physical forms of arousal than on emotional expression. The role of psychological factors in sexual arousal and responsiveness will preoccupy us often in the rest of this book. Our primary concern at this point is with the physiological mechanisms underlying these emotional processes, and we shall deal with these at the end of the chapter.  .

## Gender Differences in Sexual Arousal

Do women and men react differently to sexual stimuli? Are men "turned on" more easily than women? Are such differences, if they exist, innate or culturally determined?

Despite obvious and widespread interest in such questions, amazingly little research exists concerning differences among males and females as to what they generally consider to be erotically stimulating. Let us consider, for instance, the effect of viewing potentially prurient visual material. Kinsey reported that men are generally more stimulated than women by viewing sexually explicit materials (such as pictures of nudes, genitals, or sexual scenes) but that women are as equally stimulated as men by viewing motion pictures and reading literary material with romantic content. Most people would tend to agree to this claim on the basis of common observation. After all, the readership of "girlie" magazines, collectors of "dirty pictures," audiences at "stag films" and burlesque shows are predominantly male; and there seems to be no comparably intense interest among women.

But what about experimental rather

than reported evidence (which was the basis for Kinsey's conclusion) in this regard? A group of 50 male and 50 female students at Hamburg University matched for various background characteristics were shown sexually explicit pictures under experimental conditions.[10] In general, men did find pictures with sexual themes more stimulating when this involved isolated scenes (male subjects looking at pictures of attractive women in bikinis or exposing their genitals; female subjects looking at pictures of men in bikinis or showing naked men with erect penises). Where the scene had an interpersonal or affectionate component (kissing couple) women were equally if not more responsive. Scenes of coitus were somewhat more arousing to men, but not much more so than to women.

When looking at these pictures, the most frequent physiological reactions of the women were genital sensations (warmth, itching, pulsations), and about a fifth of the women reported vaginal moistening. Men usually responded with an erection.

In terms of the prevalence of these physiological responses, there was no significant overall difference between the sexes: 35 women and 40 men (out of 50 in each group) had some genital response. Nor was there any significant difference in the sexual aftereffects of the test: About half the subjects in each group reported increased sexual activity during the next 24 hours, involving masturbation, petting, or coitus. Increased sexual desire that was not acted on was also reported in comparable terms by both sexes.

A larger sample (128 males and 128 females) were shown films featuring male and female masturbation, petting, and coitus.[11] Once again, men responded more readily, but the difference with women was slight. There were no significant differences in the

prevalence of physiological reactions while viewing the films. Among the women 65 percent experienced genital sensations—28 percent felt vaginal moistening and 9 percent had sensations in the breast. Among men 31 percent had full and 55 percent had partial erections. About one in five men and women reported some masturbatory activity while viewing the film, and in four cases, all of them male, this was carried to the point of orgasm. During the following 24 hours there was some increase in sexual activity for both sexes, especially masturbation.

These findings tend to confirm the notion that males seem to respond more readily than females to erotic visual material. But this statement must be immediately qualified. First, the difference between the sexes is nowhere as impressive in this regard as is generally assumed. The apparent contrast is to a large extent due to social expectations and cultural convention: Women are not supposed to react as men do to explicit sexual material. This is often sufficient to inhibit their responses. In the event that their responses are not inhibited, women are likely to keep quiet about their true feelings so as not to risk social censure.

Second, whatever actual differences may exist generally between the sexes, they are superseded by the differences among the members of each sex. Thus, one will find many women who will react more positively to erotic visual stimuli than the "average" man. When this occurs, it is misleading to conclude that some women "respond like men," as if there were a fixed male standard of response. It is preferable to think of a variety of responses that men and women share in variable degree, but none of which is exclusive or characteristic of either sex.

There are at least two additional pitfalls in such comparisons. One is related to the biased selection of the erotic stimulus. Most erotic art, for instance, is produced by men. When we use these male-preferred sexual

---

[10]Sigusch, et al. (1970).
[11]Schmidt and Sigusch (1970). For a more comprehensive discussion of this research, see Chapter 7 by these same authors in Zubin and Money (1973).

representations and symbols as the standard stimuli, and women do not respond as actively as men do, it hardly follows that women are generally less responsive to erotic visual materials.

The second problem is even more fundamental and is based on sex discrepancies in the self-perception of sexual arousal. Most investigations of erotic stimulation used to rely on the verbal reports of subjects on viewing photographs of nudes and verbal descriptions or motion pictures of sexual activity.[12]

More recently, when physiological measures of arousal have been combined with verbal reports, a whole new dimension has been added to our understanding of these issues. In one study subjects listened to tape-recorded stories with erotic and romantic themes. In addition to reporting how they felt, the male subjects were monitored with a penile strain gauge and the females with a vaginal photoplethysmograph.[13] Both sexes reacted similarly in finding the tape with explicit erotic content more stimulating than those with romantic or erotically neutral content. Particularly arousing were the tapes where the female was the initiator of the encounter. There was however an interesting discrepancy between the verbal reports of the women and their physiological reactions: Only half of the women who were physiologically aroused reported this fact. The men on the other hand never failed to perceive the evidence of their physiological arousal.[14] There is no reason to believe that the female subjects were willfully concealing the fact of their being aroused. Their relative failure to per-

ceive sexual arousal must be explained therefore by anatomical differences (it is harder to miss an erection than the female counterpart such as vasocongestion) or perhaps by unconscious phychological mechanisms repressing such perception.[15]

Currently, women are becoming more willing to reveal their sexual preferences, and they tend to oppose the traditional practice of being measured by male criteria. As stereotyped views of sex-linked differences diminish, one can actually see a convergence in sexual response patterns. This has already taken place to some extent among the younger generations in the West.[16] As these changes become more generalized, we shall see what differences will remain, if any, among the responses of women and men to erotic visual stimuli.

These remarks notwithstanding, it would be premature to conclude that all currently perceived differences are due to ''sexist'' bias. There is ample evidence for discrepancies in the patterns of sexual arousal among animals. In the case of animals, communication of sexual interest predominantly occurs through physiological processes (like the secreting of pheromones) or morphological changes (such as the swelling and redness of the anogenital region of the female in estrus). Animals also engage in certain set behavioral courtship patterns. These processes are usually quite sex-specific.

What complicates matters further is the fact that such declarations of sexual interest may also serve to communicate other messages. For example, among primates rump presentation and mounting are both preludes to coitus, as well as a demonstration of submissive and dominant status respectively. To

---

[12]See, for example, Clark (1952); Byrne and Sheffield (1965); Schmidt and Sigusch (1970); Byrne et al. (1973).

[13]The penile strain gauge is a flexible loop, similar to a rubber band, that fits at the base of the penis and measures blood volume and pressure pulse, both of which reflect the degree of vasocongestion. The photoplethysmograph is an acrylic cylinder that contains a photocell and a light source. It is placed at the entrance of the vagina and detects evidence of vasocongestion.

[14]Heiman (1975).

[15]For more information on physiological measures of sexual arousal, see Zuckerman (1971), McConaghy (1967), and Heiman (1978).

[16]For references related to this trend in northwestern Europe and the United States, see Schmidt and Sigusch in Zubin and Money (1973), p. 140.

what extent such primate patterns of sexual stimulation persist among humans is not clear. The effect of hormones on libido is discussed in Chapter 4 and is obviously pertinent to the issue of sexual arousal.[17] As for other aspects of sexuality, we are just beginning to seriously view human behavior in primate perspective. There is a major shift in humans to more subtle behavioral patterns and verbal communication of sexual intent. Cultural diversity in this regard is wide, and a great deal of such behavior is barely realized consciously even by the person manifesting it.[18]

We have only dealt here so far with the question of erotic response to visual stimuli. The means of sexual arousal are so many and their underlying processes so complex that these brief references should be viewed as no more than illustrative examples. We must keep in mind that arousing sexual interest goes far deeper than mere sexual coquetry and forms the first step in the reproductive endeavor through which a species maintains itself.

## Sexual Response: General Characteristics

In response to sexual stimulation the body reacts as a whole. The components of this total response are, however, many and varied. To facilitate description we shall first outline the general behavioral characteristics of sexual response patterns and then deal with the physiological changes in the sex organs and various body systems.

We shall be dealing throughout with the ordinary patterns of human sexual response. Our descriptions are not, however, intended as standards of "normality" or "healthiness." There are innumerable variations on these patterns, and they are also perfectly "nor-

mal" and "healthy." The biological basis of sexuality does not imply that its manifestations are relentlessly uniform.

Since sexual excitement and orgasm are widely experienced phenomena, we may wonder about the need to describe them. But most of us cannot generalize from only our own experiences. There is far too much variation among individuals to permit generalization, and, furthermore, most of us are in no mood for dispassionate observation at times of sexual stimulation. There is in fact a blurring of our perceptual ability during sexual arousal, as a result of which we are not quite fully conscious of our own sensations and physiological responses.

## Approach to Orgasm

In response to effective sexual stimulation a sensation of heightened arousal develops. Thoughts and attention turn to the sexual activity at hand, and the person becomes progressively oblivious to other stimuli and events in the environment. Most people attempt to exert some control over the intensity and tempo of their mounting sexual tensions. They may try to suppress it or to ward it off by diverting their attention to other matters. Or they may deliberately enhance and prolong the feeling by dwelling on its pleasurable aspects. If circumstances are favorable to fuller expression, these erotic stirrings are difficult to ignore. On the other hand, anxiety or strong distractions may easily dissipate sexual arousal during the early stages.

Although excitement sometimes intensifies rapidly and relentlessly, it usually mounts more unevenly. In younger people the progression is steeper, whereas in older ones it tends to be more gradual. As the level of tension rises, external distractions become less effective, and orgasm is more likely to occur.

The prelude to orgasm is pleasurable in itself and can be quite satisfying. In fact, following a period of sustained excitement one may voluntarily forgo the climax. But linger-

---

[17]Also see Dmowski et al. (1974).
[18]For an anthropological review of sexual signaling, see Hewes (1973).

ing tensions usually do create irritability and restlessness if unrelieved by orgasm.

It is generally believed that men respond more rapidly to sexual stimulation and are capable of reaching orgasm more rapidly than women are. A great deal of advice in marriage manuals revolves around this very point: Because women supposedly respond more slowly, they must be stimulated for longer periods if they are to reach orgasm. This belief has a certain validity, for in clinical practice the "slower" wife usually does complain that she is "left behind." Yet there is no known physiological basis for this claimed difference, and females can respond more or less as quickly as males do to effective sexual stimulation.

The average female, for example, takes somewhat less than four minutes to reach orgasm during masturbation, whereas the average male needs between two and four minutes. Some women may achieve climax, however, in as little as 15 to 30 seconds. The disparity between the sexes in achieving coital orgasm is therefore related not to fundamental physiological differences but to the mechanical and psychological components of sexual intercourse.

The behavioral and subjective manifestations of sexual excitement vary so widely that no one description can possibly encompass them all. With mild sexual excitement, relatively few reactions may be visible to the casual observer; on the other hand, in intense excitement behavior may be quite dramatic.

The person in the grip of sexual excitement appears tense from head to toe. The activities of the muscles, though dramatic, are by no means the only sexual responses of the body. The skin becomes flushed; salivation increases; the nostrils may flare; the heart pounds; breathing grows heavy; the face is contorted and flushed; the person feels, looks, and acts quite unlike his or her ordinary self.

These phenomena may be quite mild. The results of muscular tension and vaso-congestion are inevitably present, but are not always reflected in a highly visible manner. The person may remain still or show only occasional and minimal overt responses. Movements may be deliberate and gentle. Changes in facial expression may be minor. But, no matter how attenuated his or her behavioral manifestations, the person must experience distinct increases in heart rate and heavier breathing; otherwise arousal is simply not present.

We have so far omitted all reference to changes in the sex organs, for we shall deal with them in detail later. We have also left out idiosyncratic manifestations: Some stutterers, for instance, speak more freely when sexually aroused. The gagging reflex may disappear, which explains the ability of some people to take the penis deep into their mouths. Spastics may coordinate better; those suffering from hay fever may obtain temporary relief; bleeding from cuts decreases. The perception of pain is markedly blunted during sexual arousal, which partly explains masochistic endurance of sadistic practices.

## Orgasm

Orgasm (from the Greek *orgasmos,* "to swell," "to be lustful") is one of the most intense and profoundly satisfying sensations that a person can experience. In physiological terms it consists of the explosive discharge of accumulated neuromuscular tensions. More subjectively defined, it is a high pitch of tension in which time seems to stand still for a moment. There is drive toward release and utter helplessness to stop it. In a matter of seconds it is all over, but while it lasts it seems an eternity.

The patterns of response during orgasm vary among individuals and according to age, fatigue, length of abstinence, and context. On strictly physiological grounds there is no reason why men and women should react differently during orgasm. However, psychological factors and different standards of propriety

**Box 3.2**   Varieties of Female Orgasm

The formal differentiation between types of female orgasm was originally proposed by Freud[1] and reiterated by other psychoanalysts.[2] It has thus been a tenet of orthodox psychoanalytic theory that females experience two types of orgasm: clitoral and vaginal. The term *clitoral orgasm* means orgasm attained exclusively through direct clitoral stimulation. Likewise, *vaginal orgasm* means orgasm attained through vaginal stimulation.

This dual orgasm theory assumes that in young girls the clitoris, like the male penis, is the primary site of sexual excitement and expression. With psychosexual maturity the sexual focus is said to shift from the clitoris to the vagina, so after puberty the vagina emerges as the dominant orgasmic zone. Should this transfer not occur, the woman remains incapable of experiencing vaginal orgasm and is restricted to the "immature" clitoral type.

[1] *New Introductory Lectures on Psychoanalysis* (1933), in Freud (1957–1964), Vol. XXII.
[2] For example, Ferenczi (1936), Abraham (1948), Knight (1943), Fenichel (1945), Deutsch (1945), Benedek (1952). In more current psychoanalytic literature there is a shift away from the dual orgasm hypothesis. See, for instance, Salzman (1968).

Kinsey and his associates raised doubts about the whole concept of dual orgasm. They pointed out that the vagina is a rather insensitive organ; during pelvic examinations many women simply cannot tell when the vaginal wall is being gently touched, and in surgical practice the vagina has been found to be rather insensitive to pain. Also, microscopic studies fail to reveal end organs of touch in most vaginal walls.

The data from Masters and Johnson's study supported the Kinsey point of view: Physiologically there is one and only one type of orgasm. The clitoris and the vagina respond in identical fashion, regardless of which is stimulated or, for that matter, even if neither one is directly involved (as when orgasm occurs after breast manipulation only). This finding does not mean that the subjective experience of orgasm, elicited by whatever means, is always the same. The similarity is in the physiologic manifestations only. The subjective experience of orgasm resulting from masturbation or coitus, or from coitus in a given position or with a specific person, can certainly vary tremendously. But the basis of these differences is psychological.

The initial impact of the Masters and Johnson research was so striking that the whole controversy of clitoral versus vaginal

may markedly alter the behavior of the two sexes. Differences in orgasmic response may also arise from physical considerations, such as whether the person is experiencing orgasm when lying down or standing up, and so on.

At one extreme, the overt manifestations of orgasm may be so subdued that an observer may hardly be able to detect them; on the other hand, the experience may be like an explosive convulsion. Most commonly there is a visible combination of genital and total body responses: sustained tension or mild twitching of the extremities while the rest of the body becomes rigid; a grimace or muf-

fled cry; and rhythmic throbbing of sex organs and pelvic musculature before relaxation sets in. Less commonly reactions are restricted to the genitals alone. The pelvic thrusts are followed by subdued throbbings, and general body response seems minimal.

In intense orgasm the whole body becomes rigid, the legs and feet extended, the toes curl in or flare out,[19] the abdomen becomes hard and spastic, the stiffened neck is thrust forward, the shoulders and arms are

[19] In Japanese erotic art curled toes indicate sexual excitement. The characteristic posture of the stiffened and extended feet and hands is called *carpopedal spasm*.

**Box 3.2** Continued

orgasm appeared to have been laid to rest once and for all. But this has in fact not happened and the issue continues to be debated.[3]

Part of the problem is in the difficulty of separating physiological effects from the totality of the experience.

Currently women who are aware of the Masters and Johnson findings continue to assert their perception of subjective differences, as is illustrated in Box 3.3.

The controversy actually goes beyond the matter of subjective sensation. Singer and Singer, for example, have attempted to review the issue of whether or not convulsive contractions of the vaginal introitus and the surrounding area are a necessary element in the female orgasm.[4] They argue that they are not, and instead propose that there are three types of female orgasm.

First is the "vulval orgasm," which is characterized by involuntary rhythmic contractions of the vaginal introitus ("orgasmic platform"). They claim that this is the orgasm observed by Masters and Johnson (which we shall discuss in detail in the following section). This type does not depend on the nature of sexual activity, whether it is coitus or masturbation, and has consistent physiological properties.

Second, there is the "uterine orgasm," which does not involve contractions of the vulva but is characterized by a special type of breathing: The woman takes in a series of gasping gulps of air and then momentarily stops breathing before forcefully exhaling the held breath, sometimes uttering forced sounds or incoherent words. This is followed by a feeling of relaxation and sexual satiation.

The reason that this second type of orgasm is called "uterine" is because it presumably results from the repetitive displacement of the uterus and consequent stimulation of the peritoneum (which consists of sensitive folds of tissue that cover the abdominal organs). Unlike the vulval orgasm, the uterine variety is dependent on coitus (or a close substitute), since clitoral stimulation, for example, would have no direct physical impact on the uterus.

The third type is the "blended orgasm," which, as the terms suggests, combines elements of the previous two kinds.[5]

[3]*See,* for example, Clark (1970), Robertiello (1970), Fisher (1973).
[4]Singer and Singer (1972).

[5]Also of interest in this female orgasmic controversy are the views of Sherfey (1973) and Fisher (1973).

rigid and grasping, the mouth gasps for breath, the eyes buldge and stare vacantly or shut tightly. The whole body convulses in synchrony with the genital throbs or twitches incontrollably.

At the climax the person may moan, groan, scream, or utter fragmented and meaningless phrases. In more extreme reactions there may be uncontrollable laughing, talking, crying, or frenzied movement. Such climaxes may last several minutes.

Orgasm is experienced by both men and women as intense pleasure, though its subjective components do vary somewhat between the two sexes.[20] In adult males the sensations of orgasm are linked to ejaculation,[21] which occurs in two stages. First,

[20]An interesting way of conceptualizing the orgasmic experience is to view it as an "altered state of consciousness." For an exposition of this view, see Davidson (in press).

[21]Ejaculation and orgasm are two separate processes. Orgasm can be experienced in both sexes and probably at all ages. It consists of the neuromuscular discharge of accumulated sexual tensions. Ejaculation on the other hand is experienced only by males following puberty, when the prostate and accessory glands become functional. Females do not ejaculate. The fluid that lubricates the vagina is produced during arousal and does not correspond to the male semen.

## Box 3.3   Female Experiences of Orgasm

The first of the following selections is from a Doris Lessing novel. The second description is provided by a married pair of physiologists who monitored their sexual responses in the privacy of their own bedroom:

When Ella first made love with Paul, during the first few months, what set the seal on the fact she loved him, and made it possible for her to use the word, was that she immediately experienced orgasm. Vaginal orgasm that is. And she could not have experienced it if she had not loved him. It is the orgasm that is created by the man's need for a woman, and his confidence in that need.

As time went on, he began to use mechanical means. (I look at the word mechanical—a man wouldn't use it.) Paul began to rely on manipulating her externally, on giving Ella clitoral orgasms. Very exciting. Yet there was always a part of her that resented it. Because she felt that the fact he wanted to, was an expression of his instinctive desire not to commit himself to her. She felt that without knowing it or being conscious of it (though perhaps he was

conscious of it) he was afraid of the emotion. A vaginal orgasm is emotion and nothing else, felt as emotion and expressed in sensations that are indistinguishable from emotion. The vaginal orgasm is a dissolving in a vague, dark generalised sensation like being swirled in a warm whirlpool. There are several different sorts of clitoral orgasms, and they are more powerful (that is a male word) than the vaginal orgasm. There can be a thousand thrills, sensations, etc., but there is only one real female orgasm and that is when a man, from the whole of his need and desire takes a woman and wants all her response. Everything else is a substitute and a fake, and the most inexperienced woman feels this instinctively. (From *The Golden Notebook* (London: Michael Joseph Ltd., 1962), p. 186. Reprinted by permission of Simon and Schuster and Michael Joseph Ltd. Copyright © 1962 by Doris Lessing.

This female usually experiences two orgasms during coitus, one before the male ejaculates and a second about half a minute after ejaculation commences, and these are qualitatively different. The first sometimes tends

there is a sense that ejaculation is imminent, or "coming," and that one can do nothing to stop it. Second, there is a distinct awareness of the contracting urethra, followed by fluid moving out under pressure.

Orgasm in the female starts with a feeling of momentary suspension followed by a peak of intense sensation in the clitoris, which then spreads through the pelvis. This stage varies in intensity and may also involve sensations of "falling," "opening up," or even of emitting fluid. Some women compare this stage of orgasm to mild labor pains. It is followed by a suffusion of warmth spreading from the pelvis through the rest of the body. The experience culminates in characteristic throbbing sensations in the pelvis. The female orgasm, unlike that of the male, can be interrupted.

Are all orgasms the same? The question

is impossible to answer categorically. At one level, no two experiences are ever the same because each of us is unique and because even the same person is different in some ways at different times. But for practical purposes we do consider certain experiences to be identical if in most aspects they replicate themselves.

In the latter sense the male orgasmic experience physiologically and psychologically seems to be fairly standard. There are wide differences in intensity and other aspects, as we shall discuss in connection with sexual intercourse, but all these differences are understandable as departures from a generally recognized common experience.

The situation is more complex with women. They experience the same variations as men do, but in addition there is the possibility that women experience qualitatively

**Box 3.3**   Continued

to be fairly laboured, and the second occurs almost automatically. The second is very much more intense in feeling and occurs much more reliably than the first. The first orgasm is considerably more variable and inconstant than the second, occurring sometimes soon after the beginning of coitus, sometimes after a considerable time has elapsed and sometimes not at all. During periods of heightened excitement, the first orgasm might be intentionally eliminated, or alternatively could develop into multiple orgasms, i.e. about six or seven minor orgasms, if coitus lasts long enough. This depends on the responsiveness of the penis for when the penile reactions are swift, coitus ends quickly. The most noteworthy characteristic of the first (pre-ejaculatory) orgasms is their failure to satisfy. Whether one or six occur, the sensation of incompleteness and dissatisfaction remains. These orgasms seem to serve the purpose of ensuring a satisfactory state of excitement or responsiveness for the final (post-ejaculatory) orgasm, especially if the opportunity for foreplay is limited. With the final (post-ejaculatory) orgasm complete satisfaction is obtained, irrespective of the number of orgasms which preceded it.

. . . With the final contractions of ejaculation, the male begins to relax his hold and become inert. At this point a compulsive abdominal and vaginal straining and heaving begins in the female. The penis seems to be pushed into the vaginal outlet, the shaft is gripped just below the glans by the muscles at the vaginal outlet, until the glans is made to form a tight-fitting plug in the vagina. At this stage the heaving reaches its maximum intensity and heralds the orgasmic contractions in the vagina and uterus at intervals of approximately 1 sec. The vaginal reaction to ejaculation is so automatic that it seems to have almost the quality of a reflex, and thus could, in times of stress, presumably fail. The vaginal contractions alternate with inward heaves, which carry with them a sensation of inward suction, though the intensely pleasurable orgasmic sensations are more allied to the contractions and the sense of relief these bring. The sensation of suction does not usually accompany clitoral or extra-coital orgasm. (From C. A. Fox and Beatrice Fox, "Blood Pressure and Respiratory Patterns during Human Coitus," *Journal of Reproductive Fertility,* 19:3 (August, 1969), p. 411. Reproduced by permission of Blackwell Scientific Publications Ltd.)

different orgasms that also involve differences in physiological mechanisms (*see* Boxes 3.2 and 3.3).

The theoretical fascination of this subject notwithstanding, the practical issue to bear in mind is that clitoral stimulation seems to be a crucial component in eliciting erotic responses in the majority of women. Until proved to the contrary, there would seem to be little to be gained and much to be lost to handicap women with the view that clitoral stimulation is somehow a crutch to be relied upon if "vaginal," "authentic," or some other presumably higher orgasmic response is not forthcoming.

## Aftereffects of Orgasm

Whereas the onset of orgasm is fairly distinct, its termination is more ambiguous. The rhythmic throbs of the genitals and the con-

vulsions of the body become progressively less intense and less frequent. Overwhelming neuromuscular tension gives way to profound relaxation.

The manifestations of the postorgasmic phase are the opposite of those of the preorgasmic period. The entire musculature is relaxed. The person feels an overwhelming need to rest. The head feels too heavy for the neck to support. The grasping hands and curled toes relax, and the arms and legs can be moved only with effort. The pounding heart and accelerated breathing revert to normal. Congested and swollen tissues and organs resume their usual colors and sizes. As the body rests, the mind reawakens, and the various senses regain their full acuity gradually.

The quiescent state of body and mind following sexual climax has given rise to the belief that all animals are sad following coitus.[22] Actually, for most people the predominant sensation is one of profound gratification, or peace and satiation. The contorted expression yields to one of calm. The eyes become luminous and languid, and a subtle flush lights the face.

The descent from the peak of orgasm may occur in one vertiginous sweep, or more gradually. Particularly at night the profound postcoital relaxation contributes to natural weariness, and the person may simply fall asleep. Others feel relaxed but perfectly alert or even exhilarated.

It is not unusual to feel thirsty or hungry following orgasm. A smoker may crave a cigarette. Often there is a need to urinate, sometimes to move the bowels. Idiosyncratic reactions are myriad: Some people feel numb or itch; some want more physical contact; and others want to be left alone. It is difficult to separate physiologically determined responses from psychological, or learned, patterns of behavior in this area.

Regardless of the immediate postorgasmic response, a healthy person recovers fully from the aftereffects of orgasm in a relatively short time. Protracted fatigue is often the result of activities that may have preceded or accompanied sex (drinking, lack of sleep), rather than of orgasm itself. When a person is in ill health, however, the experience itself may be more taxing.

### Orgasm in Lower Animals

The expressions of orgasm are clearly discernible among mammals. In the male animal there is no question that orgasm occurs regularly, and ejaculation marks it clearly. Female animal orgasm is more difficult to detect. In female mammals, other than humans, neuromuscular tensions do not seem to subside abruptly following coitus but rather dissipate slowly. Furthermore, female animals in heat remain responsive to sexual stimulation after coitus, rather than losing interest as males do.

Even though most infrahuman females seem to be orgasmically nonresponsive, there are records of specific instances in which female chimpanzees have seemed to exhibit all the signs of orgasm. Physiological measures of blood pressure in these animals during sexual activity also demonstrate similar elevations and depressions for both sexes, which constitute indirect evidence for the occurrence of sexual climax. The evidence from animals indicates that orgasm evolved late for females. Whereas orgasm in males is clearly the continuation of mechanisms present in infrahuman species, the human female is unique in her ability to reach orgasm so readily and unmistakably.[23]

## The Sexual Response Cycle

The physiology of sexual function has been sadly neglected in medical research. In the fourth century B.C., Aristotle observed that the testes are lifted within the scrotal sac during sexual excitement; more than 20 centuries passed before this fact was confirmed under laboratory conditions.

So far there has actually been only one extensive investigation of the physiology of orgasm, the one conducted by Masters and Johnson.[24] Their work is widely known, and numerous summaries of it are available.[25] We shall therefore outline it only briefly here.

## Research Background

Masters and Johnson were primarily interested in investigating the physiology of orgasm in a laboratory setting. Their subjects were 694 normally functioning volunteers of

---

[22]The original remark, ascribed to Galen (A.D. 130–200) was *Triste est omne animal post coitum, praeter mulierem gallumque* ("Every animal is sad after coitus except the human female and the rooster").

[23]Ford and Beach (1951), p. 30.
[24]Masters and Johnson (1966); (1979).
[25]See, for example, Brecher and Brecher (1966), Part I, and Belliveau and Richter (1970), Chapter 4.

both sexes between the ages of 18 and 89 years. These subjects thus did not constitute a random sample of the general population, and in this sense they were not "average people." Nor were they specifically selected for their sexual prowess; the only such requirement was that they be sexually responsive under laboratory conditions. In socioeconomic terms the group was as a whole better educated and more affluent than is the general population, though there was some representation across social classes.

The research procedure was to observe, monitor, and sometimes film the responses of the body as a whole and the sex organs in particular to sexual stimulation and orgasm. Both masturbation and sexual intercourse were included in the experiment. In order to observe vaginal responses, a special penislike object was used; it was made of clear plastic, which permitted direct observation and filming of the inside of the vagina.

The laboratory in which the research took place was a plain, windowless room containing a bed and monitoring and recording equipment. The subjects were first left alone to engage in sex, and only when they felt comfortable in this setting were they asked to perform in the presence of investigators and technicians monitoring the equipment (recording heart rates, blood pressures, brain waves, and so on). It was the type of setting in which hundreds of experiments of all kinds are conducted in medical centers all over the world. The only unique element was the specific physiological function under study.

During almost a decade (beginning in 1954) at least 10,000 orgasms were investigated. Because more of the subjects were women and because females registered more orgasms than males under the circumstances, about three-quarters of these orgasms were experienced by women. The same investigators have reported more recently their physiological findings in the sexual response cycle of homosexual males and females as well as those with bisexual ("ambisexual") orientation.[26]

Although no one has yet attempted to fully replicate Masters and Johnson's research, the scientific validity of their physiological findings has been widely accepted. There have been some questions raised about minor aspects of their findings, which we shall point out further on.[27]

## Response Patterns

The sexual response patterns shown in Figure 3.1 and 3.2 summarize observations by Masters and Johnson. They represent generalized patterns rather than consistent reactions of individuals. These graphic summaries do not show the common deviations from basic responses. Such deviations, however, occur mostly in the durations of phases rather than in the sequence of response in each. We can therefore view these curves as typical, while keeping in mind the wide range of variation within them.

The sexual response pattern for males (Figure 3.1) and the three patterns for females (Figure 3.2) include the same four phases: excitement, plateau, orgasm, and resolution.

These response patterns are generally independent of the type of stimulation or sexual activity that produces them. The basic physiology of orgasm is the same, regardless of whether it is brought about through masturbation, coitus, or some other activity. Differences resulting from the type of stimulation do not affect the fundamental changes manifested by the body, but may affect the intensity of responses to some extent.

We shall repeatedly emphasize the subjective experience of the fundamental similarity of sexual responses in the two sexes. There are, however, a number of differences between male and female responses that

[26]Masters and Johnson (1979).
[27]For a critique, see Fox and Fox (1969).

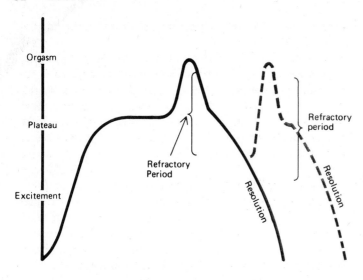

**Figure 3.1** The male sexual response cycle. From Masters and Johnson, *Human Sexual Response* (Boston: Little, Brown and Company, 1966), p. 5. Reprinted by permission.

must be noted. Some result from anatomical differences; others cannot be explained structurally and possibly reflect variations in nervous system organization.

The first major difference is in the range of variability. It is already apparent from Figures 3.1 and 3.2 that whereas a single sequence characterizes the basic male pattern, three alternatives are shown for females. Even this diagram does not fully convey the richer variety of female responses. But this does not mean that the male response is the basic norm and the female responses variant forms of the male.

The second difference between the sexes involves the presence of a refractory period in the male cycle. (A cell, tissue, or organ may not respond to a second stimulation until a certain period of time has elapsed after the preceding stimulation. This period is known as "refractory.") As is indicated in Figure 3.1, the refractory period immediately follows orgasm and extends into the resolution phase. During this period, regardless of the

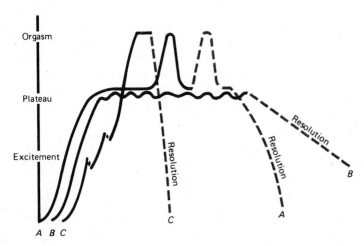

**Figure 3.2** The female sexual response cycle. From Masters and Johnson, *Human Sexual Response* (Boston: Little, Brown and Company, 1966), p. 5. Reprinted by permission.

nature and intensity of sexual stimulation, the male will not respond. He cannot achieve full erection and another orgasm. Only after the refractory period, the duration of which has not yet been specified with precision, can he do so.[28]

Females do not have such refractory periods (Figure 3.2). Even in the pattern closest to that of the male (line A), as soon as the first orgasm is over the level of excitement can mount immediately to another climax. Women can thus have multiple orgasms in rapid succession.

Men have been considered to be very limited in their capacity for multiple orgasm due to their refractory period. Among the subjects studied by Masters and Johnson only a few males seemed capable of repeated orgasm and ejaculation within minutes. More recent evidence suggests that multiple orgasm is not that rare if orgasm and ejaculation are considered separately from each other.[29] We shall return to this issue shortly when we discuss orgasm in greater detail.

Apart from these differences, the basic response patterns in the two sexes are the same. In males (Figure 3.1) and females (Figure 3.2, line A) excitement mounts in response to effective and sustained stimulation, which may be psychogenic (erotic thoughts and feelings) or somatogenic (physical stimulation), but usually involves both. Excitement may mount rapidly or more slowly, depending on various factors. If erotic stimulation is sustained, the level of excitement becomes stabilized at a high point: the plateau phase. Sometimes during this stage the point of no return is reached, and orgasm follows. This abrupt release is succeeded by a gradual dis-

persion of pent-up excitement during the resolution phase.

The lengths of these phases vary greatly, but in general the excitement and resolution phases are the longest. The plateau phase is usually relatively short, and orgasm usually takes only a minute or less. Although the diagrams do not indicate them, there may be several peaks in the plateau phase, each followed by a return to a lower level of excitement. It is also possible that excitement may not reach the point of orgasm. The overall time for one complete coital cycle may range from a few minutes to much longer.[30]

The two other female patterns (Figure 3.2, lines B and C) represent opposite alternatives to the more common pattern. In pattern B the woman attains a high level of arousal during the plateau phase but these peaks of arousal do not quite culminate in orgasmic release. Instead sexual tension is gradually dissipated through a protracted period of resolution.

In contrast, pattern C shows a more explosive orgasmic response, which bypasses the plateau phase and is followed by a more precipitate resolution of sexual tension. This form of orgasm tends to be both more intense as well as more protracted.[31]

During the various phases of the sexual response cycle numerous specific physiological changes occur, some in the genitals, some in other parts of the body. As we describe these reactions we must keep in mind that not all of these responses continue through the entire cycle. Some physiological responses (like erection of the penis) set in right away and persist throughout the entire cycle,

---

[28]The length of the refractory period seems to vary greatly among different males and with the same person on different occasions. It may last anywhere from a few minutes to several hours: The interval usually gets longer with age and with successive orgasms during a given sexual episode (Kolodny et al., 1979).

[29]Robbins and Jensen (1977).

[30]Among the Kinsey subjects, men reported reaching orgasm generally within 4 minutes, whereas it took women 10 to 20 minutes to do so. But when women relied on self-stimulation, they too could reach orgasm within 4 minutes or so.

[31]Such sustained orgasms have been labeled *status orgasmus,* or "orgasmic states," in contrast to single orgasms. This is analogous to the distinction between single convulsions and convulsive states in epilepsy.

whereas the others occur only during part of a given phase (for instance, additional coronal engorgement late in the plateau phase).

## Physiological Mechanisms of Sexual Response

Two underlying mechanisms explain how the body and its various organs respond to sexual stimulation: *vasocongestion* and *myotonia*.

Vasocongestion is the engorgement of blood vessels and increased influx of blood into tissues. Ordinarily the flow of blood through the arteries into various organs is matched by the outflow through the veins, and thus a fluctuating balance is maintained. When under various conditions blood flow into a region exceeds the capacity of the veins to drain the area, vasocongestion will result. Blood flow is primarily controlled by the smaller arteries (arterioles), whose muscular walls constrict and dilate in response to nervous impulses. The ultimate causes may be physical (for example, heat) or psychological (for example, embarrassment). Congested tissue, because of its excess blood content, becomes swollen, red, and warm. Sexual excitement is accompanied by widespread vasocongestion involving superficial and deep tissues. Its most dramatic genital manifestation is the erection of the penis. (Erection has been called the "blushing of the penis.") The response varies in different organs and at different times, but is necessarily present in sexual excitement.

Myotonia is increased muscle tension. Even when a person is completely relaxed or asleep, the muscles maintain a certain firmness or "muscle tone." From this baseline muscular tension increases during voluntary flexing or involuntary orgasmic contractions. During sexual activity myotonia is inevitable and widespread. It affects both smooth and skeletal muscles and occurs both voluntarily and involuntarily. Although evidence of myotonia is present from the start of sexual excitement, it tends to lag behind vasocongestion.

Vasocongestion and myotonia are the underlying sources of practically all physiological manifestations during sexual activity, and the reader should keep them in mind during our discussion of general and specific physiological responses.

The physiological manifestations of the sexual response cycle do not change in any fundamental way in older individuals. Aging does bring about definite anatomical and physiological alterations, but the differences in this regard between the older person and his or her former self are mainly differences in the intensity and not in the nature of the physiological response.[32]

The original findings reported by Masters and Johnson were derived from observations of orgasm attained by self-stimulation and coitus. These have now been supplemented by reports of the physiological reactions of homosexual and bisexual males and females. This evidence shows that there is no fundamental difference in sexual physiology between heterosexuals and those with homosexual orientations.[33]

## Reactions of Male Sexual Organs

### Penis

Of all the male sex organs the penis undergoes the most dramatic changes during sexual excitement and orgasm. Erection is correctly regarded as evidence of sexual excitement. Nevertheless, there are instances in which erections occur in the absence of sexual stimulation or erotic feelings and vice versa.

Erection is experienced on innumerable occasions when awake or while asleep by practically all males and may occur in earliest infancy (some baby boys have erections right after birth), as well as in old age. Erection is not an all-or-none phenom-

---

[32]Sexuality and aging, including changes in physiology, are discussed in more detail in Chapter 8.

[33]For more details, see Chapter 7 in Masters and Johnson (1979).

enon; there are many gradations between the totally flaccid penis and the maximally congested organ immediately before orgasm.

Erection occurs with remarkable rapidity. In men younger than 40 years of age, less than 10 seconds may be all the time required. There are tremendous differences among individuals, as well as among occasions for the same individual. Younger men generally respond more rapidly, but slowing down with age is not inexorable.

The excitement phase may be short, and activity may rapidly proceed to the more stable plateau phase. More often the earlier phase is quite protracted, and the varying firmness of erection reflects the waxing and waning of sexual excitement and desire. During this period a man is quite vulnerable to loss of erection. Even if sexual stimulation continues uninterrupted, distraction can cause partial or total detumescence.

During the plateau phase the penis undergoes two further though relatively minor changes. Although full erection is achieved at the end of the excitement phase, further engorgement occurs, primarily in the corona of the glans. Erection is then more stable, and the man may temporarily turn his attention away from sexual activity yet remain tumescent.

During orgasm the characteristic rhythmic contractions begin in the accessory sexual organs (prostate, seminal vesicles, and vas), but very soon extend to the penis itself (*see* Figure 3.3). Orgasmic contractions involve the entire length of the penile urethra, as well as the muscles sheathing the root of the penis. At first they occur regularly at intervals of approximately 0.8 second, but after the first several strong contractions they become weaker, irregular, and less frequent.

Ejaculation (''throwing out''), though not precisely synonymous with orgasm, is its cardinal manifestation in the adult male. The fluid, which flows in variable amounts (usually about 3 cubic centimeters, or a teaspoonful), is known as ''semen,'' ''seminal fluid,'' or ''spermatic fluid.'' It consists of sperm (which account for very little of its volume) and the secretions of the prostate (which impart to it a characteristic milkiness and odor), of the seminal vesicles, and to a much lesser extent of Cowper's glands.

Ejaculation consists of two distinct phases. During the first (emission, or first-stage orgasm) the prostate, seminal vesicles,

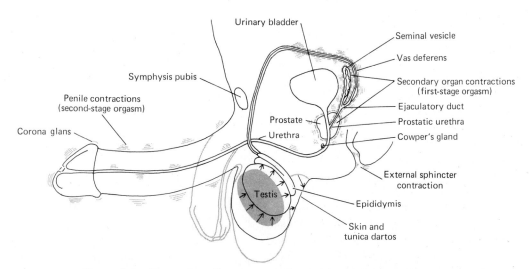

**Figure 3.3**  The male pelvis: orgasmic phase. *From Masters and Johnson, Human Sexual Response* (Boston: Little, Brown and Company, 1966), p. 184. Reprinted by permission.

and vas deferens pour their contents into the dilated urethral bulb. At this point the man feels the inevitability of ejaculation. In the second phase (ejaculation proper, or second-stage orgasm) the semen is expelled by the vigorous contractions of the muscles surrounding the root of the penis and the contractions of the genital ducts. The subjective experience at this point is one of intense pleasure associated with orgasmic throbs and the sensation of spermatic flow.

The amount of fluid and the force with which it is ejaculated are popularly associated with strength of desire, potency, and so on, but these beliefs are difficult to substantiate.

There is also a popular notion that semen is generated entirely by the testes, for during arousal the testes may feel tense and full, and this feeling is relieved by ejaculation. Nevertheless, the contribution of the testes to semen is restricted to sperm and the minimal fluid that carries them. The pelvic sensation of fullness before orgasm results from vasocongestion and sustained muscular tensions. Women have similar experiences even though they do not ejaculate.

In the resolution phase the changes of the preceding stages are reversed, and erection is lost. Detumescence occurs in two stages: First, there is a relatively rapid loss of erection, which reduces the organ to a semierect state; then there is a more gradual decongestion in which the penis returns to its unstimulated size. In general, the longer the excitement and plateau phases (and the more marked the vasocongestion process), the longer the primary stage of resolution, which in turn delays the secondary stage.

After ejaculation, if sexual stimulation continues (if the penis remains in the vagina or even if the man remains close to his sexual partner), the penis remains tumescent longer. If, on the other hand, he withdraws, is distracted, or attempts to urinate, detumescence is more rapid. A man actually cannot urinate with a fully erect penis because the internal

urinary sphincter closes reflexively during full erection to prevent intermingling of urine and semen.

### Scrotum

During the excitement phase the scrotal skin contracts and thickens with a resulting loss of the normal baggy appearance. The initial response of the scrotum to sexual excitement is thus similar to its reaction to cold (and also to fear and anger). If the excitement phase is quite prolonged, the scrotum relaxes, even though the sexual cycle is not yet completed. During the plateau and orgasmic phases there are no further changes. In the resolution phase there is usually a rapid loss of the thickening of the scrotal skin. The scrotum shows no color changes.

### Testes

The changes undergone by the testes, though not visible, are quite marked. During the excitement phase both testes are lifted up within the scrotum (*see* Figure 3.3), mainly as a result of the shortening of the spermatic cords and the contraction of the scrotal sac. During the plateau phase this elevation progresses farther until the organs are actually pressed against the body wall. Full testicular elevation is necessary for orgasm for reasons that are unclear.

The second major change is a marked increase in size (about 50 percent in most instances) because of vasocongestion. During orgasm there are no further changes.

In the resolution phase the size and position return to normal. Again the process may be rapid or slow (the pace is usually consistent for a given person) and, the longer the plateau phase, the more protracted is the process of detumescence.

### Cowper's Glands

The bulbourethral (Cowper's) glands show no evidence of activity during the excitement phase. If tension is sustained, a drop or so of clear fluid, probably produced by these glands, appears at the tip of the penis. These tiny structures are rudimentary in men. By contrast, in some animals (for example, stallions, rams, bears, and goats) these glands secrete profusely during sexual arousal.

Men vary a great deal in the production of this mucoid material. Although most secrete only a drop or none at all, a few produce enough to wet the glans or even to dribble freely. The presence of this fluid is reliable evidence of a high level of sexual tension, but its absence does not ensure the opposite. Its association with voluptuous thoughts has been well known. Medieval scholars called it the "distillate of love" and correctly distinguished it from semen. It has also been mentioned in ancient literature.[34]

It has been assumed sometimes that this secretion acts as a lubricant during coitus. Because of its usually scanty and inconsistent presence, however, this function is unlikely. A more plausible explanation is that it neutralizes the urine-contaminated acidic urethra, protecting the sperm during their passage.

Although quite unrelated to semen, this secretion sometimes contains stray sperm that have seeped out before ejaculation. Intercourse may thus result in pregnancy even though the male withdraws his penis before orgasm, which often explains unexpected pregnancies when the participants vehemently maintain that coitus stopped before intravaginal ejaculation.

### Prostate and Seminal Vesicles

The responses of the prostate gland and the seminal vesicles are similar and will therefore be described together. Overt changes are

restricted to the orgasmic phase (*see* Figure 3.3), during which they play a major role. As indicated earlier, ejaculation actually begins with the contractions of these accessory structures as they pour their secretions into the expanded urethra. The admixture of sperm from the throbbing vas deferens with the secretions of the seminal vesicles (whose walls also pulsate spasmodically) occurs in the ejaculatory duct, and sperm are thus propelled through the duct into the urethra. These accessory organs actually participate in the rhythmic convulsions of orgasm. Their contractions, along with the filling of the urethra, are responsible for the sensation that orgasm is imminent.

## Reactions of Female Sex Organs

### Vagina

In reproductive terms, the vagina corresponds functionally to the penis and the physiological responses of the vagina and penis to sexual stimulation are complementary: As the penis prepares for insertion, the vagina prepares to receive it; however, these reactions are not limited to sexual intercourse. Effective stimulation, or whatever origin, brings about the standard vaginal and penile responses.

The vagina exhibits three specific reactions during the excitement phase: lubrication, expansion of its inner end, and color change. The warm, moist vagina reveals sexual desire, and the term "lubricity" is an appropriate synonym for it. Moistening of the vaginal walls is actually the very first sign of sexual response in a woman and usually occurs within 10 to 30 seconds after erotic stimulation.[35]

The lubricatory function of the clear, slippery, and mildly scented vaginal fluid is

---

[34]The *Priapeia*, epigrams on Priapus by various Latin poets, includes the following poem (quoted in H. Ellis [1942], Vol. II, Part One, p. 153):

> You see this organ after which I'm called
> And which is my certificate, is humid;
> This moisture is not dew nor drops of rain,
> It is the outcome of sweet memory,
> Recalling thoughts of a complacent maid.

[35]However, the fact of vaginal lubrication is not always sufficient evidence by itself that a woman is ready for coitus. There are psychological factors beyond mechanical considerations to be taken into consideration.

self-evident. As the fluid is alkaline, it has also been assumed to help neutralize the vaginal canal (which normally tends to be acidic) in preparation for the transit of semen. The source of this fluid was, however, unknown until recently.

In the past it was assumed that the fluid was a female ejaculate, analogous to male semen and therefore essential to conception. This misunderstanding, which was prevalent until the seventeenth century, had some unusual repercussions. It formed the basis for theological tolerance of female masturbation if orgasm did not occur naturally during coitus; the argument was that, unless the woman complemented male semen with her own ejaculation, conception could not occur. Such a failure would negate what was considered the primary function of coitus.

Inasmuch as the vaginal wall has no secretory glands, it was assumed also that the lubricant emanated from either the cervix, the Bartholin glands, or both. It has recently been convincingly demonstrated, however, that this fluid oozes directly from the vaginal walls. The secreting mechanism is as yet unclear, but in all likelihood it is related to vasocongestion in the vaginal walls.[36]

The Bartholin glands produce a discharge, but, in analogy to the secretion of the male Cowper's glands, it tends to be scanty, erratic, and of lubricating value only in the area of the introitus, if at all. Incidentally, the clear vaginal fluid that results from sexual excitement must not be confused with the chronic vaginal discharges produced by various infections (*see* Chapter 12). Some secretions from the cervix may also be normally present, but they too do not contribute substantially to vaginal lubrication.

The second major vaginal change during the excitement phase is the lengthening and expansion of the inner two-thirds of the vagina. The ordinarily collapsed interior vaginal walls expand to create a space where the ejaculate will be deposited. The stretched walls of the vagina thus lose some of their normal wrinkled appearance.

Finally, the ordinarily purple-red vaginal walls take on a darker hue in response to sexual stimulation. This discoloration, initially patchy, eventually spreads over the entire vaginal surface. These color changes reflect the progressive vasocongestion of the vaginal walls.

During the plateau phase the focus of change shifts from the inner two-thirds to the outer one-third of the vagina. In the excitement phase the outer end of the vagina may have dilated somewhat, but in the plateau phase it becomes congested with blood and the vaginal lumen becomes at least a third narrower. These congested walls of the outer third of the vagina constitute the *orgasmic platform*. It is there that rhythmic contractions during orgasm are most evident. During the plateau phase the "tenting effect" at the inner end of the vagina progresses still farther, and full vaginal expansion is achieved. Vaginal lubrication tends to slow down, and if the excitement phase is unduly protracted, further production of vaginal fluid may cease altogether in the plateau phase.

During orgasm (*see* Figure 3.4), the most visible effects occur in the orgasmic platform. This area contracts rhythmically (initially at approximately 0.8-second intervals) from 3 to 15 times. After the first three to six contractions the movements become weaker and more widely spaced. This orgasmic pattern varies from person to person and in the same person from one orgasm to another. The more frequent and intense the contractions of the orgasmic platform, the more intense is the subjective experience of climax. At particularly high levels of excitement these rhythmic contractions are preceded by spastic (nonrhythmic) contractions of the orgasmic platform that last 2 to 4 seconds. The inner portion of the vagina does not contract but

[36]Masters and Johnson (1966), p. 70.

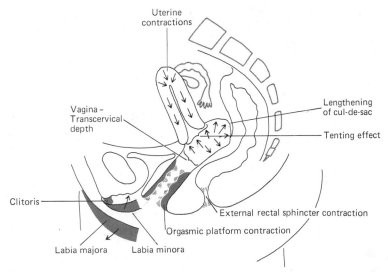

Uterine
contractions

Vagina –
Transcervical
depth

Lengthening
of cul-de-sac

Tenting effect

Clitoris

External rectal sphincter contraction

Orgasmic platform contraction

Labia majora        Labia minora

**Figure 3.4** The female pelvis: orgasmic phase. From Masters and Johnson, *Human Sexual Response* (Boston: Little, Brown and Company, 1966), p. 77. Reprinted by permission.

continues its "tenting." These observations confirm that the vagina is not a passive receptacle for the penis but an active participant in coitus.

During the resolution phase the orgasmic platform subsides rapidly. The inner walls return much more slowly to their usual form. With decongestion the color of the vaginal walls lightens over a period of 10 to 15 minutes. The process of lubrication may in rare instances continue into this phase, and such continuation indicates lingering or rekindled sexual tension. With sufficient stimulation, a second orgasm may follow rapidly.

### Clitoris

The clitoris is an exclusively sexual organ, and in contrast to its male homologue, the penis, it plays no part in reproduction and is totally independent of the urinary system. In response to sexual excitement the clitoris becomes tumescent through vasocongestion, just as the penis does. The overhanging prepuce of the clitoris does not, however, permit it to stand erect. That the clitoris does not function exactly as a miniature penis is also apparent from its relatively slow response to stimulation. The immediate counterpart to penile tumescence is vaginal lubrication, rather than clitoral vasocongestion.

The clitoris is highly sensitive. Practically all women perceive tactile stimulation in this area, and most women respond erotically to such manipulation.[37] Its shape and ability to trigger voluptuous sensations have led to comparisons with an electric bell button.[38]

During the excitement phase both the glans and the shaft of the clitoris become congested. This response is more consistent for the glans and may result in the doubling of its diameter. The vasocongestive response is more rapid and more marked if the clitoris and adjoining areas of the mons are stimulated directly. The fundamental sequence of

[37]In gynecological examinations conducted for the Kinsey study, 98 percent of the subjects were able to perceive tactile stimulation of this organ; in contrast, less than 14 percent could detect being touched in the interior of the vagina (Kinsey et al., 1953), pp. 574 and 580.

[38]" . . . a veritable, electric bell button which, being pressed or irritated, rings up the whole nervous system" (quoted by H. Ellis, 1942, Vol. II, Part One), p. 130.

changes, however, is independent of the method of stimulation.

The shape, size, and position of the clitoris in the unstimulated state and its visible tumescence during the excitement phase have no relation to the likelihood of orgasm. Important as it is for sexual stimulation, the clitoris gives no reliable clues to the subsequent course of the sexual cycle. Visible tumescence of its glans, when it does occur, coincides with the vasocongestive response of the minor labia and comes quite late in the excitement phase (when the penis has been erect for some time and the vagina is fully lubricated). The clitoral glans, once tumescent, remains so throughout the sexual cycle.

During the plateau phase the entire clitoris (glans and shaft) shows a peculiar response. It is retracted under the clitoral hood and almost disappears from view. This reaction is particularly rapid and striking in response to direct stimulation and may result in the clitoris' receding to half its unstimulated length. As the initial enlargment of the organ indicates sexual excitement, its subsequent retraction may understandably be misinterpreted as indicating loss of sexual tension. When excitement abates the clitoris reemerges from under the hood. During a protracted plateau phase there may be several repetitions of this retraction-emergence sequence, which may confuse the uninformed male who is attempting to stimulate his partner's clitoris.

During orgasm the clitoris remains hidden from view. Following orgasm it promptly (in 5 to 10 seconds) reemerges from its retracted position. The rapidity and timing of this response are comparable to the first postorgasmic loss of penile erection. Final detumescence of the clitoris, on the other hand (like the second stage of penile erection), is much slower (usually taking 5 to 10 minutes but sometimes as long as half an hour). When orgasm has not occurred, the engorgement of the glans and shaft of the clitoris may persist for hours and cause discomfort.

### Major Lips

The labia of women who have not given birth (nulliparous labia) are anatomically somewhat different from those of women who have (parous labia). These structural alterations influence the types of physiological response during the sexual cycle, especially in the major lips.

Nulliparous major lips become flattened, thinner, and more widely separated, "opening" and exposing the external genitals. This slight opening reveals the congested moist tissues between. During the plateau and orgasmic phases nulliparous major lips show no further changes. During the resolution phase they return to their decongested size and shape and resume their midline contact. Resolution proceeds rapidly if orgasm does occur. Otherwise the changes brought about during excitement take longer to dissipate. Following a protracted period of arousal, congestion may be so intense that the labia remain swollen for several hours after all sexual stimulation has ceased.

Parous major lips are larger and more pendulous and may contain permanently distended (varicose) veins. Instead of flattening, they become markedly engorged, and may double or triple in size during arousal; but they do nevertheless expose the entrance to the vaginal orifice. There are no changes during the next two phases. Resolution is more rapid if orgasm occurs. Otherwise this phase takes longer, depending on how distended the labial veins have become and how effectively they can be drained.

### Minor Lips

The alterations in the minor lips during the sexual cycle are quite impressive and remarkably consistent. As the excitement phase progresses to the plateau level, they become severely engorged and double or even triple in size in both parous and nulliparous women. These tumescent lips project between the overlying major lips and become quite apparent, which may explain the parting of the major lips during excitement.

Color changes reflect the extent of venous congestion and are therefore affected by any existing venous distension in parous women. During the plateau phase the minor lips become progressively pink, or even bright red in light-complexioned women. In parous women the resulting color is a more intense red or a deeper wine color.

This vivid coloration of the minor lips has been observed so consistently that they have been called the "sex skin" of the sexually excited woman.[39] If erotic stimulation continues beyond this point, orgasm is inevitable; but if stimulation is interrupted, orgasm will not occur. Orgasm does not occur unless labial congestion first reaches this peak. In this sense the "sex skin" is comparable to the full testicular elevation of the male: Both herald impending orgasm.

### Bartholin's Glands

Bartholin's glands respond to sexual stimulation by secreting a few drops of mucoid material rather late in the excitement phase or even in the plateau phase. They appear to be most effectively stimulated by the action of the copulating penis over a long period of time, and parous women have a more generous production of the secretion. The contribution of these glands to vaginal lubrication or to neutralizing the acidic vaginal canal is relatively minor.

### Uterus

Despite its being hidden from view the uterus has long been known to participate actively in the changes of the sexual response cycle. It has often been assumed that the contractions of the uterus during coitus cause the semen to be sucked into its cavity. Masters and Johnson found no evidence to support this assumption.[40] One also encounters references in literature to a woman's enjoyment resulting from the penis ramming against the uterine cervix. This too is totally unsubstantiated. In fact, the cervix is remarkably insensitive; it can even be cut without pain.

The uterus responds to sexual stimulation initially by elevation from its usual position. (This reaction does not occur if the uterus is not resting in its normal anteverted position.) This reaction pulls the cervix up and contributes to the tenting effect in the vagina. Full uterine elevation is achieved during the plateau phase and is maintained until resolution, when the organ returns to its usual position over a period of 5 to 10 minutes.

In addition, the uterus clearly shows the effects of the two main physiological phenomena: vasocongestion and myotonia. The former is manifested in a distinct increase in size during the earlier phases, which returns to normal following orgasm; the latter is apparent in the activity of the uterine musculature, culminating in distinct contractions.

Orgasmic contractions start in the fundus and spread downward. Although they occur simultaneously with those of the orgasmic platform, they are less distinct and more irregular. The cervix shows no specific change until the resolution phase, when the external cervical opening may dilate to some extent immediately after orgasm. The more intense the orgasm, the greater is the likelihood of this cervical reaction. Because of inevitable changes in the cervix during child-

---

[39]Masters and Johnson (1966), p. 41.

[40]This is contradicted by Fox et al. (1970), who found that a pressure gradient exists between the vagina and uterus immediately following female orgasm. Their proposed explanations for the reason why Masters and Johnson failed to find evidence for uterine suction are in Fox and Fox (1967).

birth, this reaction is best seen in the nulliparous woman.

## Extragenital Reactions

As was indicated earlier, the responses of the body to sexual stimulation are not restricted to the sex organs. As extragenital manifestations are quite similar in the two sexes, we shall describe them together.

### Breasts

Even though male breasts also respond to sexual stimulation, changes during the sexual cycle are far more striking in the female. Our description, therefore, refers primarily to the latter.

Erection of the nipple is the first response of the breast to sexual stimulation. It occurs in the excitement phase and is the result of the contraction of "involuntary" muscle fibers rather than vascular congestion. Engorgement of blood vessels is, however, responsible for the enlargement of the breasts as a whole, including the areolae.

In the plateau phase the engorgement of the areolae is more marked. As a result, the nipples appear relatively smaller. The breast as a whole expands farther during this phase, particularly if it has never been suckled (it may increase by as much as a fourth of its unstimulated size); a breast that has been suckled may change little in size or not at all. During orgasm the breasts show no further changes. In the resolution phase, along with the rapidly disappearing sexual flush, the areolae become detumescent, and the nipples regain their fully erect appearance ("false erection"). Gradually breasts and nipples return to normal size.

Changes in the male breast are inconsistent and restricted to nipple erection during the late excitement and plateau phases. Male nipples are rarely stimulated directly during heterosexual activity (sometimes they are during homosexual contact), but the nipple

reactions nevertheless appear in more than half the instances observed.

### Skin

The significance of skin changes accompanying emotional states is well known. We blush in embarrassment, flush in anger, turn pale in fear. These surface reflections of inner feelings are manifest in the infusion or draining of blood from the vessels in the skin and are controlled by the autonomic nervous system. It is thus hardly surprising that sexual activity results in definite skin reactions, consisting of flushing, temperature change, and perspiration.

The flushing response is more common in women. It appears as a discoloration, like a rash, in the center of the lower chest (epigastrium) during the transition from the excitement to the plateau phase. It then spreads to the breasts, the rest of the chest well, and the neck. This sexual flush reaches its peak in the late plateau phase and is an important component of the excited, straining, and uniquely expressive physiognomy of the woman about to experience the release of orgasm. During the resolution phase the sexual flush disappears very quickly; the order reverses its spread so that discoloration leaves the chest last.

Temperature changes during the sexual response cycle have not yet been measured. Although there is no evidence that the temperature of the body as a whole changes, people do frequently report feelings of pervasive warmth following orgasm; and there are popular references to sexual excitement as a "glow," "fever," or "fire." Superficial vasocongestion is the likely explanation of this sensation.

Perspiration (apart from that caused by physical exertion) occurs fairly frequently during the resolution phase. Among men this response is less consistent and may involve only the soles of the feet and the palms of the hands. There may be a great deal of physical activity during a sexual encounter and, when the atmosphere is warm, sweating is greatly increased and may occur throughout the sexual cycle. Perspiration is one of the means by which the overheated body cools itself.

### Cardiovascular System

Just as the heart races and pounds during fear, anger, and excitement, it also responds to sexual stimulation by beating faster. This reaction is usually not immediate; and mild, transient erotic thoughts may not alter the heart rate. But significant levels of sexual tension and certainly orgasm do not occur without some elevation of the pulse rate.

In the plateau phase the heart rate rises to 100 to 160 beats a minute (the normal resting heart rate is 60 to 80 beats a minute). The blood pressure also registers definite increases parallel to that in the heart rate. These changes are quite significant, and are comparable to levels reached by athletes exerting maximum effort. They entail considerable strain on the cardiovascular system, which is easily handled most of the time; people with heart disease, however, require medical guidance in this regard.

### Respiratory System

Respiration and heart rate are interrelated through complex physiological mechanisms, so they respond concurrently to demands on the body. The most common example is physical exercise. Changes in respiratory rate lag behind those in heart rate. Faster and deeper breathing becomes apparent in the plateau phase, and during orgasm the respiratory rate may go up to 40 a minute (the normal rate is about 15 a minute, inhalation and exhalation counting as one). Breathing, however, becomes irregular during orgasm, when the individual may momentarily hold his breath and then breath rapidly. Following orgasm he may take a long deep breath or sigh as he sinks into the resolution phase. Along with the pulse and blood pressure, respiration returns to normal rate and depth.

The flaring nostrils, heaving chest, and gasping mouth that accompany sexual experience are popularly known and caricatured in the stylized panting of comedians to suggest erotic excitement. Some of the panting and grunts uttered during orgasm result from involuntary contractions of the respiratory muscles, which force air through the spastic respiratory passages. Following orgasm the soft palate relaxes, and the person may make snoring noises.

The changes manifested by the cardiovascular and respiratory systems are partly caused by muscular exertion and are nonspecific to sex. Apart from these changes, however, are some that occur in response specifically to sexual stimulation. Changes in facial expression and gasping for breath during orgasm raise the possibility that the individual suffers a lack of oxygen (anoxia), but this surmise has not been conclusively documented.[41]

### Digestive System

The response of the digestive tract to sexual stimulation can best be observed at its beginning and end: the mouth and the anus. During sexual arousal the secretion of saliva increases and the person may literally drool. During intense erotic kissing (or mouth-genital contact) increased salivation is very apparent.

The anus is a sensitive area, which, apart from its proximity to the genitals, appears to be intimately involved in both eliciting and responding to sexual stimuli. Some people react to anal stimulation erotically, whereas others are indifferent or disgusted. Anal stimulation is not an exclusively male homosexual practice; it occurs in heterosexual relations also.

[41]For a review and discussion of blood pressure and respiratory patterns during human coitus, see Fox and Fox (1969).

## Box 3.4  Reactions of Sex Organs during the Sexual Response Cycle

| Male | Female |
|---|---|
| *Excitement Phase* | |
| Penile erection (within 3–8 seconds)<br>*As phase is prolonged:*<br>  Thickening, flattening, and elevation of scrotal sac<br>*As phase is prolonged:*<br>  Partial testicular elevation and size increase | Vaginal lubrication (within 10–30 seconds)<br>*As phase is prolonged:*<br>  Thickening of vaginal walls and labia<br>*As phase is prolonged:*<br>  Expansion of inner ⅔ of vagina and elevation of cervix and corpus<br>*As phase is prolonged:*<br>  Tumescence of clitoris |
| *Plateau Phase* | |
| Increase in penile coronal circumference and testicular tumescence (50–100% enlarged)<br>Full testicular elevation and rotation (orgasm inevitable)<br>Purple hue on corona of penis (inconsistent, even if orgasm is to ensure)<br>Mucoid secretion from Cowper's gland | Orgasmic platform in outer ⅓ of vagina<br>Full expansion of ⅔ of vagina, uterine and cervical elevation<br>"Sex-skin": discoloration of minor labia (constant, if orgasm is to ensue)<br>Mucoid secretion from Bartholin's gland<br>Withdrawal of clitoris |
| *Orgasmic Phase* | |
| *Ejaculation*<br>Contractions of accessory organs of reproduction: vas deferens, seminal vesicles, ejaculatory duct, prostate<br>Relaxation of external bladder sphincter<br>Contractions of penile urethra at 0.8 second intervals for 3–4 contractions (slowing thereafter for 2–4 more contractions)<br>Anal sphincter contractions (2–4 contractions at 0.8 second intervals) | *Pelvic response (no ejaculation)*<br>Contractions of uterus from fundus toward lower uterine segment<br>Minimal relaxation of external cervical opening<br>Contractions of orgasmic platform at 0.8-second intervals for 5–12 contractions (slowing thereafter for 3–6 more contractions)<br>External rectal sphincter contractions (2–4 contractions at 0.8-second intervals)<br>External urethral sphincter contractions (2–3 contractions at irregular intervals, 10–15% of subjects) |
| *Resolution Phase* | |
| Refractory period with rapid loss of pelvic vasocongestion<br>Loss of penile erection in primary (rapid) and secondary (slow) stages | Ready return to orgasm with retarded loss of pelvic vasocongestion<br>Loss of "sex-skin" color and orgasmic platform in primary (rapid) stage<br>Remainder of pelvic vasocongestion as secondary (slow) stage<br>Loss of clitoral tumescence and return to position |

**Box 3.5**    General Body Reactions during the Sexual Response Cycle

| Male | Female |
|---|---|
| *Excitement Phase* | |
| Nipple erection (30%) | Nipple erection (consistent) |
|  | Sex-tension flush (25%) |
| *Plateau Phase* | |
| Sex-tension flush (25%) | Sex-tension flush (75%) |
| Carpopedal spasm | Carpopedal spasm |
| Generalized skeletal muscle tension | Generalized skeletal muscle tension |
| Hyperventilation | Hyperventilation |
| Tachycardia (100–160 beats per minute) | Tachycardia (100–160 beats per minute) |
| *Orgasmic Phase* | |
| Specific skeletal muscle contractions | Specific skeletal muscle contractions |
| Hyperventilation | Hyperventilation |
| Tachycardia (100–180 beats per minute) | Tachycardia (110–180 beats per minute) |
| *Resolution Phase* | |
| Sweating reaction (30–40%) | Sweating reaction (30–40%) |
| Hyperventilation | Hyperventilation |
| Tachycardia (150–180 beats per minute) | Tachycardia (150–180 beats per minute) |

Stimulation of the anus has well-known repercussions in the body: Stretching the anal sphincter induces inhalation, but contraction makes exhalation difficult. The rhythmic contraction of the anus, along with the flexing of the buttocks, induces sexual tension. Some women are able to reach orgasm through this maneuver alone.

Sexual activity elicits anal responses (noted by Aristotle). During the excitement and plateau phases, the rectal sphincter contracts irregularly in response to direct stimulation. The more striking reactions, however, occur during orgasm, when involuntary contractions can be seen to occur at approximately the same 0.8-second intervals as do the throbs of the orgasmic platform and the penile urethra. Anal contractions, which do not always occur, usually involve only two or four spasms. The rectal sphincter relaxes while the manifestations of orgasm are still in progress elsewhere.

**Urinary System**

The male urethra is an integral part of the penis, and its changes during the sexual response cycle have already been described in that connection. In some women the urethra undergoes a few irregular contractions during orgasm. Unlike the anal spasms these contractions are asynchronous and quite feeble. The urge to urinate after orgasm has already been mentioned.

The multiplicity of changes involving various males and females organs makes is difficult to maintain an overall view of the progression of events during each phase of the sexual response cycle. Boxes 3.4 and 3.5 are intended to highlight the temporal interrelations of the reactions of different body parts and should convey an impression of the orderly yet variable progression of events.[42]

[42]The tables are adapted from Beach, ed. (1965), pp. 517 and 522. A more detailed set of tables can be found in Masters and Johnson (1966), pp. 286–293.

## Neurophysiological Basis of Sexual Functions

Most of the information presented so far in this chapter has been descriptive. Underlying the patterns of sexual stimulation and response that we have considered, there exist complicated mechanisms of control. Some of these are hormonal in nature and will be discussed in the next chapter. Others are neurophysiological and, though closely integrated with endocrine functions, can be discussed separately for our purposes here.

All sensory input must ultimately be interpreted in the brain before a particular sensation is experienced. But, apart from this process, the brain can use memory and imagination to initiate sexual excitement without external sensory stimuli. Thus every form of behavior we observe has a corresponding set of events that silently unfold in the nervous system. No thought, however trivial, and no emotion, however ethereal, can exist in an empty skull. Therefore all sexual experience must be finally understood at the level of neurophysiological activity. This is not to say that the neurophysiological component is the most "important" or that such understanding in itself will be an adequate substitute for other ways of conceptualizing human experience, such as in psychological or philosophical terms. But neither can we ever dispense with the physical basis of these mental events if we aspire to comprehend them fully.

Unfortunately we have only a rudimentary knowledge at this time of the neuroendocrinological aspects of sexual functions and behavior. Until recent years virtually nothing was known about the parts of the brain involved in the control of even such basic functions as erection and ejaculation.[43] During the past several decades a good deal of research has changed this. But much of this work is still in an experimental stage. Because of this and the fact that even an elementary understanding of these matters would require more background in neuroanatomy and physiology than can be reasonably expected, we shall not attempt to discuss this area in any depth here. Enough will be said, however, to illustrate some of the basic neurophysiological processes at work in sexual functions and point out the awesome complexity of what remains to be discovered.[44]

Our primary interest here is in the neural control of sexual function. Sexual activity entails a great deal more than just genital function and involves virtually every system of the body. A great many of the components of sexual activity are thus specific or exclusive to sexual functions. For instance, touch receptors (Meissner's corpuscles) embedded in the skin merely inform the brain that a certain area of the skin is being stimulated; it is up to the brain to interpret this touch as a lover's caress.

There are special nerve endings and end organs that respond specifically to cold, warmth, pain, touch, and the like; but none that is specialized to respond selectively to sexual stimulation. The nerve endings in the glans penis are thus no different from those in the fingertips. Also, sensory messages are transmitted from the body to the brain and spinal cord, and responses are transmitted to the body parts through the same networks of nerves regardless of whether or not the activity is sexual. In this sense nerves, like telephone cables, are ignorant of the content of the messages they carry.

The nervous control of sexual functions is coordinated at two levels: in the spinal cord and in the brain. Normally, control at these levels is closely coordinated, but some functions like erection and ejaculation can also occur independently of the brain.

[43]MacLean (1976).

[44]For a review of animal studies in neural control of sexual behavior see Bermant and Davidson (1974). Brain mechanisms of sexual functions are discussed in MacLean (1976) and Whalen (1977). For a broader introduction to neuropsychology see Thompson (1967) and Pribram (1977).

## Spinal Control of Sexual Function

Sexual functions are controlled at their most elemental level by spinal reflexes. Reflexes are the basic units of nervous organization and at their simplest have three components: a receptor, a conductor or transmitter, and an effector. The receptor can be any nerve that detects and conveys sensory information (such as touch, pressure, or pain) to the transmitter neurons in the spinal cord, which interpret the sensory input and convey the appropriate response to the third component, the end organs ("effectors"), thus completing the reflex arc. The end organ may be a muscle that contracts or a gland that secretes in response to such stimulation. An example of a simple reflex is withdrawal of the hand after touching a hot object. Another well-known example is the knee-jerk response to tapping the patellar tendon. Reflexes are involuntary in the sense that their response is automatic and does not require a "decision" by the brain. The brain is conscious of these responses and may be able to inhibit the reflexive response to a variable extent.

The examples of the reflex acts given above involve the sensory motor system. There are other reflexes, including those involved in sexual functions, that belong to the autonomic part of the nervous system. This system operates involuntarily to control many of the internal functions of the body. Its activities are normally carried out below the level of consciousness.

The autonomic nervous system has two main subdivisions, both of which are involved in sexual function: the parasympathetic and the sympathetic. Stimulation of one or inhibition of the other has the same overall effects.

One of the basic functions of autonomic nerves is to control the flow of blood by constricting or dilating arteries. In the case of genital blood vessels the effect of parasympathetic stimulation is arterial dilation and the effect of sympathetic stimulation is arterial

construction. The reflexive processes in sexual functions have been better elucidated in the male. Hence we shall first discuss penile erection and ejaculation.

## Mechanism of Erection and Ejaculation

The three major spinal components that control erection and ejaculation are easier to comprehend when considered separately. The first component in the process of erection consists of the sensory (afferent) nerves, which convey a variety of stimuli from the genitalia to the spinal cord. As shown in Figure 3.5 there are two nerves that transmit such stimuli to the sacral portion of the spinal cord (segments S2 to S4): The _pudendal nerve_ carries impulses elicited by the stimulation of the surface of the penis, and the _pelvic nerve_ transmits impulses elicited by pressure and tension deep within the penile structures (_corpora cavernosa_ and _corpus spongiosum_).

The second division of the reflex mechanism consists of the spinal cord centers for erection and ejaculation (Figure 3.6). As described above, the pudendal and pelvic nerves convey sensory impulses to the S2 to S4 seg-

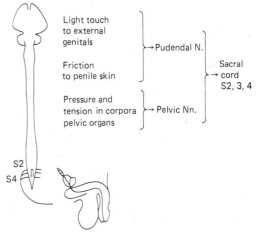

**Figure 3.5** Peripheral afferent stimuli. From Tarabuley, "Sexual Function in the Normal and in Paraplegia." _Paraplegia 10_ (1972), 202. By permission.

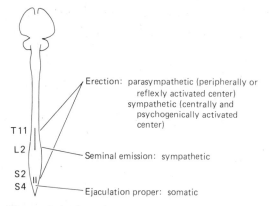

Erection: parasympathetic (peripherally or reflexly activated center)
sympathetic (centrally and psychogenically activated center)

Seminal emission: sympathetic

Ejaculation proper: somatic

**Figure 3.6**   Spinal cord centers. From Tarabuley, "Sexual Function in the Normal and in Paraplegia." *Paraplegia 10* (1972), 204. By permission.

ments of the spinal cord, which contain the primary parasympathetic "erection center," activated reflexly by external or peripheral stimulation. A second spinal center located higher up (segments T11 to L2) is part of the sympathetic system, which also plays a role in the process of erection. This center is thought to mediate psychogenically or centrally induced stimulation from the brain and results in erection. This apparently contradictory situation in which parasympathetic and

sympathetic stimulation seem to produce the same reaction will be clarified shortly.

The spinal ejaculatory centers also have dual locations. The main autonomic center is in the sympathetic segment of the spinal cord (T11 to L2). This is where the first phase of orgasm, seminal emission, is triggered. The second component, ejaculation proper, is controlled by the sacral portion of the cord (S2 to S4), but this time not by its autonomic component but rather by the somatic or voluntary part.

The third division of the reflex arcs, consisting of the effector (or efferent) nerves, is shown in Figure 3.7. Note that the same nerves (pelvic and pudendal) that were shown in Figure 3.5 to transmit impulses to the spinal cord also convey impulses from the cord to the tissues. This is because what appears as a single nerve has a large number of component nerves that, like a cluster of telephone cables, communicate different messages to and fro.

The efferent or outgoing nerves to the tissues carry the messages that result in the final outcomes of erection and ejaculation. These involve the parasympathetic, sympathetic, and somatic parts of the nervous sys-

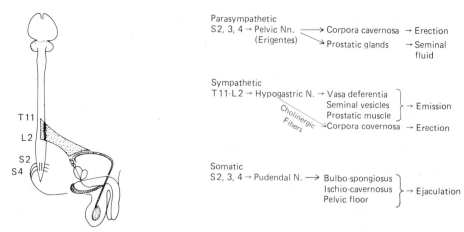

Parasympathetic
S2, 3, 4 → Pelvic Nn. ——→ Corpora cavernosa → Erection
(Erigentes)  ↘ Prostatic glands   → Seminal fluid

Sympathetic
T11-L2 → Hypogastric N. → Vasa deferentia ⎫
                          Seminal vesicles ⎬ → Emission
            *Cholinergic Fibers*  Prostatic muscle ⎭
                          ↘ Corpora covernosa → Erection

Somatic
S2, 3, 4 → Pudendal N. ——→ Bulbo-spongiosus ⎫
                          Ischio-cavernosus ⎬ → Ejaculation
                          Pelvic floor ⎭

**Figure 3.7**   Peripheral efferents. From Tarabuley, "Sexual Function in the Normal and in Paraplegia." *Paraplegia 10* (1972), 204. By permission.

tem, all of which work in a closely integrated fashion. The parasymphatic system is primarily responsible for erection (but also for the stimulation of the prostate gland, which discharges seminal fluid). The sympathetic system is primarily responsible for ejaculation in concert with the somatic component (but also is involved in the process of erection). To clarify these variously overlapping and sequential activities, we shall next consider what transpires physiologically during an entire sexual orgasmic cycle.

When stimulated by appropriate incoming sensations the parasympathetic reflex erection center in the spinal cord responds by sending out impulses through parasympathetic nerves to the arteries conveying blood to the spongy tissue of the penis. Ordinarily cavernous and spongy bodies receive modest amounts of blood from the arteries, which are then drained by the veins. Under parasympathetic stimulation these arteries dilate. With the onrush of blood, the penile tissues rapidly fill up as the veins of the penis, hampered by their valves and the compression of their thin walls by the swelling organ, cannot handle the outflow of blood. The penis is thus rapidly engorged so that it stiffens and stands erect. Loss of erection results from a reversal of this process: Sympathetic nerve fibers constrict the arteries, cutting down the inflow of blood, and drainage through the veins increases until the organ returns to a flaccid condition.

To understand how sympathetic stimulation can also produce erection we need to digress briefly to consider further how these two divisions of the autonomic system operate. The major differences between the effects of these two systems depend on the type of hormones secreted by their nerve cells (neurons). In the case of the parasympathetic system the chemical secreted is acetylcholine, and therefore these neurons are called *cholinergic*. The sympathetic neurons secrete norepinephrine (also called noradrenaline) and are therefore said to be *adrenergic*. Al-

though this distinction generally holds, some sympathetic neurons are cholinergic and therefore their effect mimics that of parasympathetic stimulation. As shown in Figure 3.7, the sympathetic fibers (in the hypogastric nerve) that act on the corpora cavernosa are cholinergic; hence their effect, similar to parasympathetic stimulation, is to promote vasocongestion and thereby result in erection.

When sexual stimulation is effective to the point of triggering orgasm then another reflexive mechanism comes into play. The first phase of orgasm, emission, is triggered by the sympathetic center (T11–L2) through the sympathetic fibersin the hypogastric nerve. This results in the contraction of the vas deferens, seminal vesicles, and prostate gland, and the emptying of their contents into the urethra. In the next phase, impulses through somatic nerves to the muscles surrounding the penis (bulbospongiosus, ischiocavernosus, and other pelvic muscles) result in their contraction and the ejaculation of semen.

Emission and ejaculation are thus two closely integrated yet separate physiological processes. Ordinarily they occur together, but under some conditions one or the other may be experienced separately. Thus emission may take place without ejaculatory contractions, in which case semen simply seeps out of the urethra. Or ejaculatory contractions occur without having been preceded by the emission of semen.

The mere fact that there is no expulsion of semen during ejaculation does not necessarily mean that emission has not occurred. It is possible for ejaculation to be retrograde. That is, instead of the ejaculate being forced out of the urethra, it may be pushed in the opposite direction and empty into the urinary bladder.[45]

The processes described so far are the reflexive aspects of penile erection. They are

[45]For further discussion of the spinal sexual mechanism, see Tarabulsi (1972), Weiss (1973), Hart (1978), and Davidson (in press).

independent of the brain, in the sense that they can occur without assistance from higher centers. For instance, a man whose spinal cord has been severed above the level of the spinal reflex center may still be capable of erection. He will not "feel" the stimulation of his penis; for that matter, he may even be totally unconscious. But his penis will respond "blindly" as it were.

The independence of reflex centers does not mean that they cannot be influenced by the brain, however. Intricate networks link the brain to the reflex center in the spinal cord. Purely mental activity may thus trigger the mechanism of erection without physical stimulation, or it may inhibit erection despite persistent physical stimulation. Usually the two components operate simultaneously and complement each other. Spurred by erotic thoughts, the man initiates physical stimulation of the genitals; conversely, physical excitement inspires erotic thoughts.

The instances in which erection seems to be nonsexual in origin involve tension of the pelvic muscles (as when lifting a heavy weight or straining during defecation). Irritation of the glans or a full bladder may have the same effect. Erections that occur in infancy are explained on a reflex basis also. An additional gruesome example is erection experienced by some men during execution by hanging.

## Reflexive Mechanisms in Women

It is generally assumed that there are spinal centers in women that correspond to the erection and ejaculatory centers of men. The reflexive centers in the female spinal cord are less well identified in part because the manifestations of orgasm are relatively more difficult to ascertain among female experimental animals.

The vasocongestive response that underlies male erection results in comparable tumescent changes in the clitoris and the rest of the genital tissues as we have seen. Likewise, the orgasmic response is similar, al-though there are some important differences as was discussed earlier.

The vasocongestion response, which results in the swelling of genital tissues and vaginal lubrication, is under parasympathetic control. The orgasmic phase in women does not have an emission component and therefore corresponds to the second or ejaculatory phase of orgasm of the male consisting of the rhythmic contractions of the musculature of the genital organs and the muscles surrounding them. Women of course do not ejaculate in the sense of emitting some counterpart of semen.[46]

## Brain Mechanisms

We have made repeated references to the influence of the brain on sexual function without specifying how this works and which portions of the brain are involved. We have avoided this issue so far because relatively little is known about the representation of sexual functions in the brain, and what is known is so complex that it is difficult to deal with it adequately within the confines of this book.

It is important to realize at least that we already know enough about the sexual functions of the brain to go beyond the usual generalities that the brain is the "most important erogenous zone." The following examples will indicate the nature of the research done in this field and some of its salient findings.

During the past several decades, many neurophysiologists and psychologists have been investigating the neurophysiology of sexual functions and emotional states using experimental animals as subjects.

There are three general investigative approaches to the study of sexual neurophysiological mechanisms in the brain. First is the use of electrical stimulation, whereby erection, ejaculation, and copulatory behavior are elicited by activating various areas in the

---

[46]For a contradictory view claiming that women do ejaculate, see Sevely and Bennett (1978).

## Box 3.6   The Klüver-Bucy Syndrome

In 1937 two investigators reported that removal of the temporal lobes of the brain in the monkey resulted in striking behavioral changes, including increased sexual activity involving autoerotic, heterosexual, and homosexual behavior.[1]

This phenomenon, now named after the two investigators, Klüver and Bucy, has been reproduced in man by bilateral removal of the temporal lobes. The change in sexual behavior was manifested as follows:

. . . Fifteen days after the operation the patient's attention was attracted by the sexual organs of an anatomic scheme hanging on the wall of the examining room. On that occasion he displayed to the doctor, with satisfaction, that he had spontaneous erections followed by masturbation and orgasm. From then on the patient gradually became exhibitionistic. He wanted to show his sexual organ erect to all doctors. He never manifested any sexual aggressiveness toward persons of the female sex for whom, on the contrary, he showed indifference in contrast with his behavior before the operation. Homosexual tendencies, clearly expressed by verbal invitations to some doctors, were soon noticed. Although monotonously insistent, he did not manifest the slightest aggressiveness either verbally or with gestures. At present, about two years after the operation, the patient is now in a mental hospital, practices self-abuse several times a day, but shows no aggressiveness either toward the male or female sex. The homosexual tendency still persists.[2]

Further evidence implicating the temporal lobe in sexual function comes from the observation of hypersexual episodes in temporal lobe epilepsy when sites in the temporal lobes cause the abnormal stimulation.[3]

[1]Klüver and Bucy (1937).
[2]Terzian and Calle-Ore (1955).
[3]Blumer (1970).

brain. The second approach is through destruction of specific brain areas that affect sexual behavior. It is possible for instance to eliminate copulation in male rats, cats, dogs, and rhesus monkeys by medical prioptic lesions. Similarly, sexual behavior may be eliminated in female animals through hypothalamic lesions.[47] If stimulation of a brain area results in a given sexual reaction, such as erection, and destruction of the region eliminates such response, then the inference is that the brain area in question is at least one link in the chain of brain mechanisms controlling that function.

A complementary approach is to identify brain areas that have an inhibitory effect on a given function. How we act is the outcome of the "push" to behave in a certain way as well as the inhibitory "pull" restraining us from doing so. The effect of inhibitory

mechanisms becomes manifest by the activities that appear or are modified as a result of the removal or destruction of the parts of the brain exerting the inhibitory effect. An example involving the temporal lobes is provided in Box 3.6.

Many of the sites that seem to be linked to sexual function (including parts of the temporal lobes) are part of a set of structures deep within the brain that are referred to as the *limbic system*. These parts of the brain belong to the phylogenetically older part of the brain in contrast to the cerebral hemispheres, which are newer structures in an evolutionary sense. The limbic system is not a single anatomical entity. Rather, as is shown in Figure 3.8, it has a number of discrete components with numerous pathways to other centers in the brain. But taken as a whole, these structures constitute an integrated network. The components of the limbic system are arranged in the form of a ring, one in

[47]Davidson (in press).

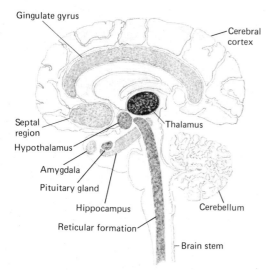

**Figure 3.8** The limbic system of the human brain. The thalamus, hypothalamus, pituitary gland, and the reticular formation are interrelated with the limbic system but are not a part of it. From Harlow, et al., *Psychology*. San Francisco: Albion Publishing Company, 1971, p. 156. Reproduced by permission.

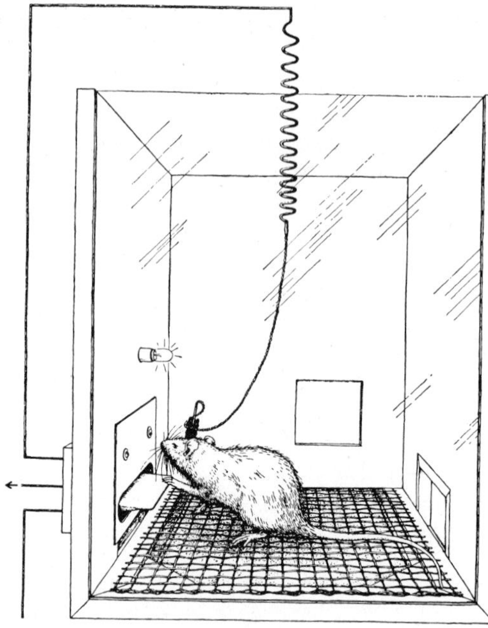

**Figure 3.9** Self-stimulation circuit allowing the rat to excite "pleasure centers" in the brain electrically by pressing on a treadle. From James Olds, "Pleasure Centers in the Brain." Copyright © 1956 by *Scientific American, Inc.* All rights reserved.

each cerebral hemisphere. Together, they surround the brain stem (which is the continuation of the spinal cord into the brain) and the paired cluster of nuclei called the thalamus.[48]

Stimulation of various parts of the limbic formations elicits alimentary, agressive, defensive, as well as sexual responses. For example, stimulation of the septal region, among other parts, elicits penile erection, mounting, and grooming in the male animal. However, the limbic formations are not the only ones involved in sexual behavior. For instance, excitation of thalamic sites results in seminal emissions and ejaculatory sites have been located along the spinothalamic tract.[49] It is also now possible to monitor the activity

of brain waves during orgasm (*see* Box 3.7).

So far we have dealt with the more specific components of the brain mediating sexual functions. How, through all this, does the brain experience pleasure? An interesting research approach has led to the location of "pleasure centers" in the brain. This research was initiated by Olds, working with rats.[50] Microelectrodes were implanted in the brain of the animal and attached to a circuit allowing the rat to stimulate its brain electrically by pressing on a treadle (*see* Figure 3.9) Ordinarily, rats under such an experimental setup will spontaneously press the lever several times in an hour. But when the electrodes were placed in certain portions of the brain (located in the hypothalamus, thalamus, and mesencephalom), the experimental rat would go on pressing the lever as often as 5000

[48]The term "limbic" is derived from the Latin for "fringe" or "border."

[49]For further discussion of brain mechanisms, see Chapter 4 by MacLean in Zubin and Money (1973), and MacLean (1976).

[50]Olds (1956).

## Box 3.7    Brain Wave Manifestations of Orgasm

The electroencephalogram (EEG) is the record of the electrical activity of the brain. The record reflects the voltage fluctuations between two points on the scalp. The EEG

From H. D. Cohen, R. C. Rosen, and L. Goldstein, "Electroencephalographic Laterality Changes during Human Sexual Orgasm." *Archives of Sexual Behavior* 5:3 (1976), 189–199.

tracings below are from the right and left parietal channels of a male subject. They successively show brain activity prior to sexual arousal (baseline), 30 seconds prior to orgasm, during orgasm, and 2 minutes following orgasm. Note that the change in amplitude in brain waves is more marked in the right hemisphere than in the left.

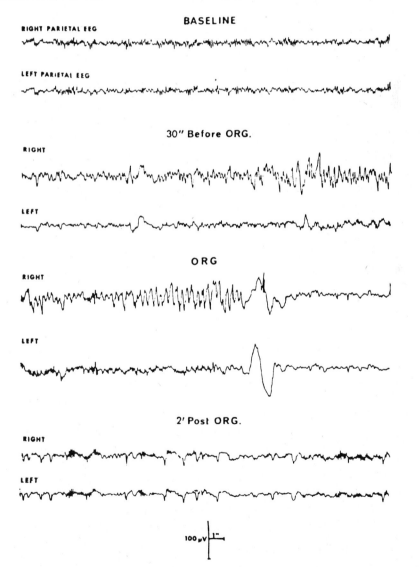

BASELINE

RIGHT PARIETAL EEG

LEFT PARIETAL EEG

30" Before ORG.

RIGHT

LEFT

ORG

RIGHT

LEFT

2' Post ORG.

RIGHT

LEFT

100 μV

times an hour, and would do this despite hunger and thirst until it was exhausted.

Since the stimulation of these locations appeared to be so rewarding, Olds called them "pleasure centers." Curiously enough, the same areas or perhaps systems inexorably interwoven with the pleasure mechanism turned out to be highly unpleasant when stimulated, so that the rat would avoid exciting them as vehemently as it would persevere in stimulating the rewarding parts. In addition, definite pathways in the brain are involved in the sensation of pain as such.

More recently, the presence of pleasure and pain systems has been shown among humans. The subjects were two patients with electrodes implanted in their brains for therapeutic purposes (one was a psychiatric male patient, the other an epileptic woman). The areas that yielded pleasure or stimulation corresponded roughly to the same regions in the animal brains. Pleasure was clearly of a sexual nature when the septal region or the amygdala of the limbic system was stimulated. Furthermore, whenever the patient was sexually aroused, brain wave changes emanating from the septal region could be detected. Such arousal could be elicited by external erotic stimuli (pictures, films), by having the subject fantasize, through the direct delivery of chemicals to the septal region, or through the administration of euphoria-producing drugs. Arousal could also be accomplished by electrical stimulation carried out by either the experimenter or the patient.

In addition to verbally reporting feelings of sexual arousal and appearing to be aroused, one of the patients, just like the experimental rat, stimulated himself incessantly (1500 times over three hours on one occasion) and begged to stimulate himself a few more times whenever the stimulating apparatus was to be taken away from him.[51] However, there is as yet no drug that when taken orally or by injection will reliably stimulate these parts of the brain and produce pleasurable sensations.[52]

These pleasure centers in man are in areas that lie close to the regions where stimulation leads to erection. They are also interconnected with other thalamic areas which receive sensory input from the body surface. One can thus begin to see the various bits and pieces fitting together to provide a neuroanatomical basis which would explain how incoming erotic tactile stimuli would activate sexual behavior and reinforce it with pleasurable feelings.

Our discussion so far has focused on various types of brain centers. Although the localization of physiological functions in the brain provides very valuable information, the final answers to how the brain regulates sexual functions is not likely to come from the identification of "sex centers" but from a broader understanding of the integrated brain systems that are involved with sexuality and other related processes. As Beach has stated, "Instead of depending on one or more 'centers,' sexual responsiveness and performance are served by a net of neural subsystems including components from the cerebral cortex down to the sacral cord. Different subsystems act in concert, but tend to mediate different units or elements in the normally integrated patterns."[53]

Another major consideration to bear in mind is that the nervous and endocrine systems work in close coordination in regulating sexual functions. The hypothalamus, for instance, is an important "headquarters" for both neural and endocrine control of sexual activity. The discussion of hormones in the chapter that follows is thus a logical succession to our consideration of the physiology of sexual function.

---

[51]For further discussion of this work see Heath (1972).

[52]Cox (1979).
[53]Beach (1977), p. 216.

# Chapter 3

# Sex Hormones and Reproductive Biology

Sex hormones play a very important role in reproductive biology. The internal and external differences in anatomy between males and females are a direct result of the effects of sex hormones. X or Y sex chromosomes program the sex glands to produce female or male sex hormones, and these hormones lead to the development of female or male reproductive organs in the fetus. At puberty the sex hormones are responsible for the maturation of these organs as well as the development of additional anatomical features like breast enlargement in females and enlargement of the larynx (voice box) in males. Sex hormones also play an essential role in all aspects of reproductive physiology, including ovulation and pregnancy in females and sperm production in males.

Although the effects of sex hormones on anatomy and physiology in humans are well established, their role in human sexual behavior is not clear. In most animals sex hormones play a critical role in mediating both sexual and aggressive behavior. Removal of the sex glands in animals leads to the rather striking differences in behavior seen, for example, between a bull and its castrated counterpart, a steer. A spayed cat shows little interest in sex, and a chicken given shots of male sex hormones will quickly rise to the top of the pecking order. There are no comparably dramatic or predictable behavioral effects of sex hormones to be found in humans.

Although the effects of sex hormones on the human brain are not well understood, important discoveries in recent years have proved that the brain regulates the production of sex hormones by the sex glands. These discoveries have provided new insights into such phenomena as infertility caused by emotional stress. In this chapter we shall discuss the role of sex hormones in reproductive biology and, more cautiously, we shall describe some possible effects of sex hormones on human behavior.

## Basic Endocrinology

A brief review of what hormones are and how they work should facilitate the reader's understanding of the topics covered later in this chapter. *Endocrinology* is the study of the secretions of endocrine, or *ductless,* glands. In contrast to some glands (like the sebaceous or salivary glands) the endocrine glands secrete their products directly into the bloodstream. Hormones are chemical compounds produced by the endocrine glands; they exert profound physiological effects on specific tissues or organs to which they are transported by the bloodstream. The endocrine glands include such structures as the thyroid, parathyroid, adrenal, and pancreas glands. For our purposes, however, we shall concentrate on the sex glands (the ovaries and testes), the *pituitary* gland, and the *hypothalamus.*

## The Pituitary

The pituitary gland (known also as the *hypophysis*) is the most complex of all endocrine glands. It is a pea-sized structure located at the base of the brain and connected to a part of the brain called the hypothalamus by a system of microscopic blood vessels and nerve fibers (*see* Figure 4.1). A variety of hormones are produced by the pituitary, and their chemical structures resemble protein molecules. The actions of the pituitary hormones include stimulating other endocrine glands to produce their hormones and stimulating the growth and maturation of various tissues.

Two pituitary hormones, called *gonadotrophins,* are of particular interest because they stimulate the sex glands: the *follicle-stimulating hormone* (FSH) and the *luteinizing hormone* (LH). A third pituitary hormone, *prolactin,* stimulates milk production in the female breast and is discussed in Chapter 5. In the female, FSH and LH stimulate the ovaries to manufacture and secrete the female sex hormones, *estrogen* and *progesterone.* In the male, LH is usually called *interstitial-cell-stimulating hormone* (ICSH) because it stimulates the interstitial (Leydig's) cells of the testes to manufacture and secrete the male sex hormone, *testosterone.* All the sex hormones belong to a group of chemical substances called *steroids,* which resemble in structure (but not in activity) that well-publicized chemical culprit, cholesterol. Steroid hormones are widely used compounds in medicine. Birth control pills consist of mixtures of synthetic female sex steroids. The drug cortisone, identical to a steroid hormone manufactured by the adrenal glands, is used for the treatment of a wide variety of ailments from arthritis to poison oak.

The early history of endocrinology is essentially the history of the discovery of the effects and the chemistry of steroid hormones. Classical methods of studying hormone activity involve depriving the organism of the hormone and observing the changes that occur; then administering doses of the hormone to demonstrate reversal of the effects noted during deprivation.

The effects of testosterone deprivation were first recorded among castrated males in ancient Egypt, China, and elsewhere. Aristotle noted the effects of castration on men and birds in the fourth century B.C. Castration of a rooster prevents growth of the cock's comb; in 1849 Berthold showed that this effect could be reversed by transplanting testes from another rooster to a castrated one. In 1889 the French physiologist Charles Brown-Sequard claimed to have experienced increased potency after treating himself with a testicular extract. This, and other highly publicized dramatic effects (longevity, youthful appearance, energy, and virility) attributed to extracts of the sex glands, tended to give endocrinology a somewhat disreputable flavor among the more conservative members of the medical profession.

In recent years, however, there has been a resurgence of interest in the study of hormones and, in particular, the mechanisms by which the brain can influence the timing and extent of hormone secretions. The study of these mechanisms is now a separate field called *neuroendocrinology.* Two anatomical sites in the head are of primary interest to neuroendocrinologists, the pituitary gland and the *hypothalamus.* Direct nerve connections from the brain to the *neurohypophysis* (posterior portion of the pituitary) have been known to exist for some time, but no nerve fibers have direct connections with the *adenohypophysis* (anterior portion of the pituitary). Since the sex gland stimulating hormones (*gonadotrophins*)[1] are produced

---

[1]Also called gonadotropins.

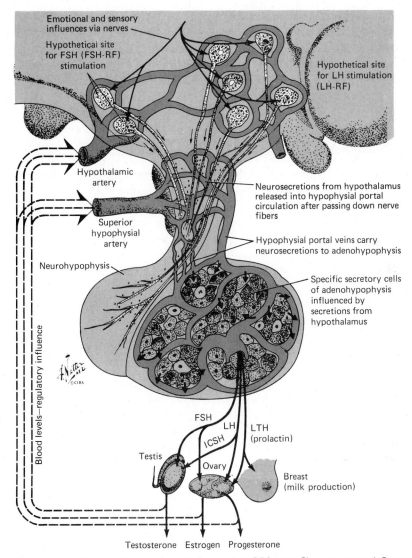

**Figure 4.1**   The pituitary gland. © Copyright 1953, CIBA Pharmaceutical Company, Division of CIBA–GEIGY Corporation. Reproduced, with permission, from The Ciba Collection of Medical Illustrations by Frank H. Netter, M.D. All rights reserved.

exclusively by cells in the adenohypophysis, the anatomical facts seemed to dictate a model in which the anterior pituitary and the sex glands are an autonomous system operating on a simple feedback principle; that is, gonadotrophins are released until the sex glands produce sufficient sex hormones to "turn off" the pituitary gonadotrophin secre-

tion. The pituitary monitors the level of sex hormones passing through it in the bloodstream, much as a thermostat monitors the temperature in the surrounding air and turns off the furnace when the temperature reaches the thermostat setting. This model is consistent with numerous observations in animals, where increasing the amount of circulating sex hormones (for example, by injection) decreases pituitary gonadotrophin output and, conversely, where decreasing the sex hormones (for example, by castration) stimulates gonadotrophin output. However, other observations do not fit this simple feedback model.

The cyclic secretions of hormones during the menstrual cycle indicate rhythmic variations in the messages being sent via the bloodstream from pituitary to ovaries. Furthermore, physical and psychological stress, which are perceived and appreciated via the brain and nervous system, have long been known to alter pituitary-dependent events, such as menstruation. The implication seems clear—the pituitary is subject to regulation by the brain, just as the thermostat in the feedback model is subject to changes in setting based on "outside" information (for instance, one might raise the temperature setting in anticipation of unusually cold weather, and so forth).

## The Hypothalamus

It has now been firmly established that the part of the brain that controls the pituitary is the hypothalamus. Located just above the pituitary, the hypothalamus receives sensory inputs from virtually every other part of the central nervous system. The hypothalamus is very small in relation to the rest of the brain, weighing about 5 grams and comprising only about 1/300 of the whole brain. (The small size of this important part of the brain is one reason why research has been difficult and painstaking.) Some of the first experiments that demonstrated the significance of the hypothalamus involved the stimulation or destruction of discrete groups of cells in various animals. It was shown that selective stimulation or ablation in specific areas of the hypothalamus will cause alterations of estrus cycles, lactation (milk production), and a number of other bodily functions such as temperature regulation, growth, and sleep.

Although initially puzzled by the absence of nerve connections between the hypothalamus and the anterior part of the pituitary (which produces gonadotrophins), researchers began to focus on the *portal* system of blood vessels linking the pituitary and hypothalamus (*see* Figure 4.1).[2] By the 1960s the generally accepted theory of neuroendocrinology was that the hypothalamus controlled the anterior pituitary by means of substances transmitted through the portal blood vessels, rather than by nerve impulses. Proof of the theory involved isolating these substances, purifying and identifying them, and demonstrating that they in fact stimulated the pituitary when administered experimentally. By the 1970s several hypothalamic hormones (also called "releasing factors") had been isolated, identified, and shown to be pituitary stimulants. In 1977 the Nobel prize in medicine was awarded to three scientists who were responsible for much of this research.[3]

The hypothalamic hormones are short-chain polypeptides, resembling fragments of larger protein molecules in chemical structure. The two factors of greatest significance for reproductive functions are known as the follicle-stimulating hormone releasing factor (FSH-RF) and the luteinizing hormone releasing factor (LH-RF).[4] As the terms imply, these brain-generated chemicals stimulate the release of FSH and LH from the pituitary gland. Hypothalamic hormones are extremely potent chemicals and are present in

[2]These vessels were first identified by a Hungarian pathologist in 1927, who noted that they were enlarged in people who died suddenly and violently. Their significance was not realized until many years later.

[3]A. V. Schally, R. S. Yalow, and R. Guillemin. For a detailed review of this work see Schally (1978).

[4]There is a trend toward using the term "releasing hormone" (RH) rather than "releasing factor" (RF). Thus, FSH-RF and FSH-RH are synonyms, as are LH-RF and LH-RH. It has been suggested that FSH-RF and LH-RF are identical compounds (Schally, 1978) but this view is disputed by others.

only minuscule amounts in the brain.[5] Research using naturally occurring extracts is therefore slow, but the possibility of synthetic versions of these hormones is on the horizon. Synthetic hypothalamic hormones would not only facilitate research but also clinical applications to problems like infertility.

## Neuroendocrine Control Mechanisms

We have already introduced the concept of feedback in the context of the thermostat model of pituitary function. This basic concept is still valid in many respects, but recent discoveries have shown that control of hypothalamic and pituitary hormone secretions is more complex than was previously thought to be the case. Furthermore, the hypothalamus is now viewed as the "ultimate" regulator, rather than the pituitary.[6] Figure 4.2 is a diagrammatic representation of current understanding of neuroendocrine control mechanisms, using LH-RF and LH as examples.

The hypothalamus receives information from two general sources: the nervous system and the circulatory system. Information from the environment as well as internally generated information (including, perhaps, emotions like anxiety) reach the hypothalamus via the nervous system. Information regarding the levels of various hormones in the body reaches the hypothalamus via the bloodstream (including, perhaps, the specialized network of vessels called the portal system). The hypothalamus responds to these inputs by either increasing or decreasing its production of hormones (releasing factors). In the

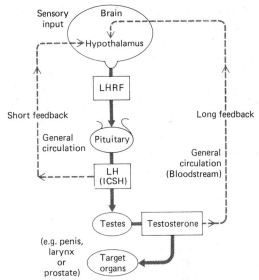

**Figure 4.2**   Diagrammatic representation of the neuroendocrine feedback mechanisms. LH-RF, LH, and testosterone are used as examples.

case shown in Figure 4.2, the hypothalamus sends LH-RF to the pituitary via the portal system. The pituitary responds by releasing LH (also known as ICSH) into the bloodstream. LH reaches the testes and stimulates the release of the male sex hormone, testosterone, which travels (also via the bloodstream) to various target organs. The increased level of testosterone in the bloodstream reaches the hypothalamus (and pituitary) via the general circulation and, in effect, conveys the information that the order for more testosterone has been received and obeyed. LH-RF and LH production then slows or stops until the testosterone levels start to drop off, signaling a need for more LH-RF (and LH). This loop between the hypothalamus and the testes is called the *long feedback* loop and is probably the primary control mechanism for sex hormone production. There is some evidence for a *short feedback* loop between the pituitary and hypothalamus via the bloodstream whereby increased LH levels

---

[5]It takes millions of animal hypothalami to yield a few milligrams of LH-RH. (The hypothalami are removed from animal brains supplied by meat packing companies.) In humans, doses of 25 micrograms (25 millionths of a gram) of LH-RH have been shown to be physiologically effective. (Schally, 1978).

[6]Until the past decade the pituitary was often referred to as the "conductor of the endocrine orchestra." It is now referred to as the "concertmaster" (Tepperman, 1973, p. 36).

"turn off" LH-RF production even before testosterone gets there.[7]

Two general observations should be clear at this point. First, to the extent that normal reproductive biology depends upon sex hormones, more is required than normal sex glands. (In fact, as we shall see shortly, conditions like precocious puberty are often caused by disorders of the pituitary or hypothalamus, not the sex glands.) Second, it is possible to "fool" the hypothalamus and pituitary by taking hormones in pill form. This raises the blood level of circulating hormones and turns off the production of LH-RF or FSH-RF (depending on which hormones are ingested). This is the principle involved in birth control pills (*see* Chapter 6).

## Female Reproductive Endocrinology

### Puberty

A girl's first menstrual period (menarche) occurs several years after the beginning of the physical changes that define puberty. Puberty begins somewhere between the ages of 9 and 12 for most girls. Menarche usually occurs between the ages of 11 and 14. Reasons for the timing of these events are not well understood (*see* Box 4.1).

#### External Changes

The most obvious change in the outward appearance of a girl during puberty is the development of the rounded contours that distinguish the adult female profile from that of the male. This process begins with an increase in secretion of FSH by the pituitary gland.

FSH stimulates the ovaries to produce

[7]There is suggestive evidence to indicate the possibility of an *ultrashort feedback* loop, which involves direct reflux of LH back to the hypothalamus via the portal system. For more detailed information on endocrinology see Reichlin et al. (1978), Ganong and Martin (1978), Tepperman (1973), and Austin and Short, vol. 3 (1972).

estrogen (a collective noun referring to a group of chemically related female sex hormones produced in the ovaries). The estrogen, in turn, travels through the bloodstream to the breasts, where it stimulates growth of breast tissue. (This and other hormonal pathways at puberty are summarized in Figure 4.3.) The pigmented area around the nipples (the areola) becomes elevated, and the breasts begin to swell as the result of development of ducts in the nipple area and an increase in fatty tissue, connective tissue, and blood vessels. The milk-producing part of the breast (*mammary gland*) does not develop fully during puberty and indeed does not become fully mature and functional until childbirth.

At the same time that estrogen is inducing the development of fatty and supporting tissue in the breasts of the young girl, a similar process is occurring in the hips and buttocks. These changes may be more pronounced in one region than in another in a particular woman. In some cultures the size and shape of the buttocks are important indices of femininity. In others greater emphasis is placed upon the breasts.

Another visible change at puberty is the appearance of *pubic* and *axillary* (underarm) *hair*. This hair is coarser and has more pigment than does that of the scalp, and its growth is stimulated partly by estrogen and partly by hormones secreted by the adrenal glands, also in response to a stimulating hormone (adrenocorticotrophic hormone, or ACTH) from the pituitary gland.

The adrenal glands are paired glands situated just above each kidney. In the female they produce small amounts of *androgens* (male sex hormones), which stimulate growth of pubic and axillary hair and are perhaps related to the female sex drive. In addition, the adrenal glands produce cortisone, adrenaline, and other substances. Excessive secretion by the adrenal glands, a pathological condition known as "Cushing's disease," may result in excessive growth of facial hair.

## Box 4.1    The Onset of Puberty

Puberty refers to the maturation of the reproductive organs and associated changes in body structure and appearance. One of the greatest unsolved mysteries in developmental biology is the question of what triggers the events of puberty. The sex glands are capable of producing the required sex hormones much earlier, and yet they do not (with rare exceptions). Attention has focused on the hypothalamus and the pituitary, since their production of stimulating hormones has been found to increase at the onset of puberty. The result is increased production of sex hormones and the resultant developments described in this chapter.

But the next question posed by this evidence is what triggers this increased stimulation on the part of the hypothalamus (and pituitary)? Current theory is that the hypothalamus gradually becomes less sensitive to the small amounts of sex hormones in the bloodstream during childhood. In feedback terms, it takes larger amounts of sex hormones to "turn off" the hypothalamic "gonadostat," with the result that the hypothalamus sends messages to the pituitary to increase FSH and LH production until, finally, adult levels of sex hormones are circulating in the bloodstream (Grumbach et al., 1974). Even if this theory is correct, it does not explain why puberty occurs when it does. One possible explanation is based on evidence that achievement of a certain critical weight (rather than age) triggers the events of puberty (Frisch, 1974). This evidence helps to explain the phenomenon of puberty occurring at earlier ages during the last century, since children now are bigger sooner. This theory and evidence have been criticized on various grounds, however, and are not yet generally accepted (Falkner, 1975).

In brief, we understand the sequence of events and the role of hormones during puberty, but we still do not know for certain what triggers the onset of puberty. Given the amount of learning that must take place before humans are capable of becoming competent parents, it does make sense for the reproductive system to mature late in development, whatever the mechanism.

A frequent source of embarrassment at puberty is the appearance of facial *acne*, a condition that is generally transient and seems to be related to changes in hormones occurring at this time of life. Although it generally presents no more than a cosmetic problem, acne may sometimes be severe and require medical treatment, particularly if accompanied by infection of the skin, which can lead to permanent scars. These scars can sometimes be satisfactorily removed later by various plastic surgery techniques, which should be performed by a competent physician. Acne tends to be more severe in girls whose menstrual cycles are very irregular.[8]

[8]Lloyd (1964), p. 184.

Estrogen also causes accelerated growth of the external genitals at puberty, including enlargement of the labia and associated structures. The clitoris enlarges under the stimulation of androgens from the adrenal glands. At the same time there is a noticeable increase in skeletal size, which is related to increased secretion of the growth hormone by the pituitary gland.

### Internal Changes

Estrogen specifically stimulates growth of the uterus and vagina during puberty. The muscular wall of the uterus enlarges and its glandular lining also develops. The lining of the vagina is extremely sensitive to estrogen, and its thickness is proportional to the amount

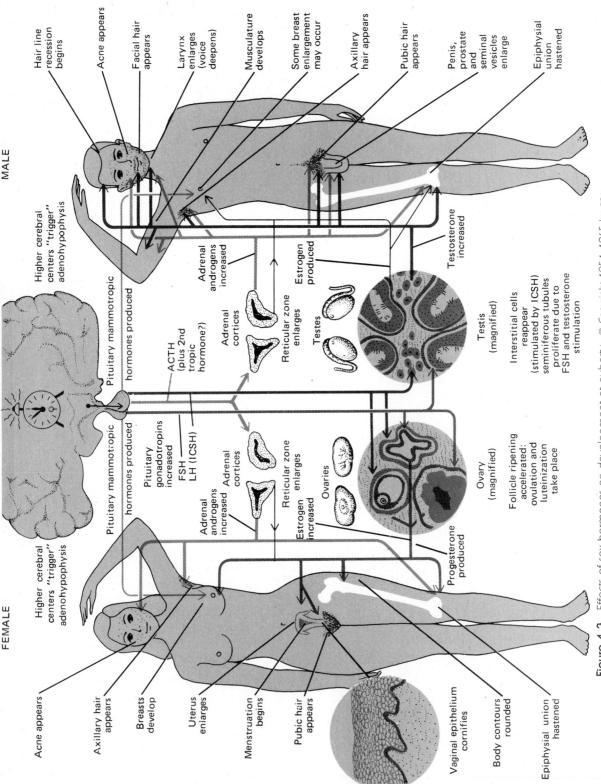

**Figure 4.3** Effects of sex hormones on development at puberty. © Copyright 1954, 1965 by CIBA Pharmaceutical Company, Division of CIBA-GEIGY Corporation. Reproduced, with permission, from *The CIBA Collection of Medical Illustrations* by Frank H. Netter, M.D. All rights reserved.

MALE

FEMALE

Hair line recession begins
Acne appears
Facial hair appears
Larynx enlarges (voice deepens)
Musculature develops
Some breast enlargement may occur
Axillary hair appears
Pubic hair appears
Penis, prostate and seminal vesicles enlarge
Epiphysial union hastened

Higher cerebral centers "trigger" adenohypophysis

Pituitary mammotropic

Pituitary mammotropic hormones produced

ACTH (plus 2nd tropic hormone?)

Adrenal cortices

Reticular zone enlarges

Adrenal androgens increased

Estrogen produced

Testes

Testis (magnified)

Interstitial cells reappear (stimulated by ICSH) seminiferous tubules proliferate due to FSH and testosterone stimulation

Testosterone increased

Higher cerebral centers "trigger" adenohypophysis

Pituitary mammotropic hormones produced

Pituitary gonadotropins increased
FSH
LH (ICSH)

Adrenal androgens increased

Adrenal cortices

Reticular zone enlarges

Estrogen increased

Ovaries

Ovary (magnified)

Follicle ripening accelerated: ovulation and luteinization take place

Progesterone produced

Acne appears
Axillary hair appears
Breasts develop
Uterus enlarges
Menstruation begins
Pubic hair appears
Vaginal epithelium cornifies
Body contours rounded
Epiphysial union hastened

of this hormone present at any given time. Examination of a mucus smear containing cells from the vaginal walls is a simple and clinically useful test for determining how much estrogen is present. As the vaginal wall matures, the pH of the secretions that moisten its surface changes from alkaline to acid.[9]

Under the influence of estrogen the female pelvis enlarges during puberty and assumes contours different from those of the male pelvis. Ultimately, estrogens also prevent further growth of the long bones of the skeleton (counteracting the effects of the growth hormone), and usually no further increase in height occurs after about age 17 in girls. An estrogen deficiency in late puberty may cause great height in a girl, for estrogen normally applies brakes to skeletal growth. Many other factors are also involved in determining the height of a given individual.

Estrogen secretion increases in quantity during puberty, mediating the various external and internal physical changes that we have just described. Moreover, it takes on a cyclical pattern of secretion, which results in certain cyclical phenomena in the female, the most obvious of which is the menstrual cycle.

### Becoming an Adult

In many societies menarche has been regarded as the time when a girl becomes a woman, and girls often marry shortly after the first menstruation. Puberty rites often occur at the time of the first menstruation. These rites are sometimes only family affairs, but in other societies the ritual is quite elaborate. Among some California Indian tribes, for instance, a girl was segregated from the rest of the tribe during her first menstrual period. She lived in a special hut built for this purpose and had to eat and bathe in a prescribed manner, using special implements. The Chiricahua Apaches also had elaborate ceremonies at menarche and believed that at this time girls possessed certain supernatural powers, including the power to heal and to bring prosperity.

In U.S. culture puberty does not mark the entrance into adulthood. Nevertheless, girls are quite sensitive in responding to the physical maturity of others in the peer group. Studies have shown, unexpectedly, that girls who mature late score highest on peer-group ratings of popularity and prestige.[10]

An interesting but unexplained observation is that the average age of menarche in Western countries has been gradually declining. In 1860, for example, a girl usually had her first period between 16 and 17 years of age, whereas in 1960 it usually occurred between 12 and 13. This phenomenon is usually attributed to better nutrition, but other factors may also be involved. Animal studies do not entirely support the notion that better nutrition alone hastens the beginning of puberty. In any event, recent evidence shows that the trend toward earlier menarche is now leveling off in Western countries.[11]

## The Menstrual Cycle

*Menstruation* (from the Latin word for "monthly") is the periodic uterine bleeding that accompanies the ovarian cycle and the shedding of the lining of the uterus. Menstruation is regulated by hormones and is seen only in female humans, apes, and some monkeys. An ovarian, or estrus, cycle occurs in other mammals but is not accompanied by bleeding. The length of the ovarian cycle is specific for each species. It is approximately 28 days in the human, 36 days in the chimpanzee, 20 days in the cow, 16 days in sheep, and five days in mice. Dogs and cats usually ovulate only twice a year.

[9]The pH scale is a commonly used index of acidity and alkalinity based on hydrogen ion concentration. The scale runs from 0 to 14; the neutral point is 7. Values from 7 to 14 indicate an alkaline state, from 7 to 0 an acid state.

[10]Hamburg and Lunde (1966), p. 3.
[11]Marshall (1977), p. 70.

## Box 4.2    Sex Differences in Physical Ability

Many people believe that males are naturally superior to females in athletics due to biological advantages in strength and endurance. The fact that 70 to 80 percent of world-class athletes in strength events use anabolic steroids (which are similar to male sex hormones) further reinforces the idea that testosterone (a male sex hormone) gives men a major advantage over women in athletics. Numerous studies on average males and females appear to confirm these ideas. Before puberty girls and boys of the same age have almost identical strength and endurance. However, after puberty, most males improve significantly in physical ability while most females either make only minor gains or actually lose strength and endurance.

In the past two decades, however, exercise physiologists have begun to study woman athletes, and a new view of female athletic ability has emerged.[1] With adequate physical training, women can develop considerable athletic ability. It has only been in recent years that many

[1]Wilmore (1975, 1977) and Mathews and Fox (1976).

females have been encouraged to develop their athletic abilities. Most girls have been discouraged from pursuing vigorous physical activity once they reach menarche. They have been pushed toward a sedentary life-style in which their physical fitness can only deteriorate. As women have become more serious competitors in the world of sports, they have begun to close the gap that has long separated female and male athletes, at least in some sports. For example, the Olympic records of women's and men's times in the 400-meter free style swimming events reveal that the men were 16 percent faster than the women in 1924, 11 percent faster in 1948, and only 7 percent faster in 1972. Both women and men have been improving their times throughout this period, but the women have been improving faster. The female record in 1970 was faster than the male record in the mid-1950s.

When comparing highly trained female and male athletes, there are many indications that women may have the same potential for strength and endurance that men do. Part of the difference in

---

The first menstruation occurs usually between the ages of 11 and 14 years, as has been noted, and may come as a shock to the girl who has not been prepared for it. For the first few years after the menarche the girl's menstrual periods tend to be irregular, and ovulation does not occur in each cycle. For some time after her first period, then, the girl is relatively infertile, but she can nevertheless become pregnant.

For the first few years after menarche a young woman's menstrual cycles may be very irregular in length, but by age 18 or 20 her periods usually assume a certain rhythm. Although there is considerable variation among

mature women in the frequency of menstrual periods, most cycles fall into the range of 21 to 35 days, with a mean of about 28 days.

The occurrence of irregular menstrual cycles in women after the age of 20 may be related to various factors, some of which are not well understood. Certainly, psychological states are important, particularly a prolonged or severe emotional stress. It was not uncommon for women to cease menstruating while imprisoned in German concentration camps during World War II (often before malnutrition or physical illness had developed). Gynecologists also report that some women cease menstruating or menstruate irregularly while

**Box 4.2**    continued

female and male performances results from the fact that males are larger and heavier than females. When one compares the performance of well-trained females and males of the same weight, many differences disappear. The strength of the lower extremities is virtually identical when comparing women and men athletes of similar weight. Females' legs are actually 5.8 percent stronger than males' legs when comparing people with the same lean body weight (i.e., total weight minus fat). This indicates that women's muscles are not inferior to men's. It is true that men show a distinct superiority in upper body strength. Well-trained female athletes are only 30 to 50 percent as strong in the upper body as well-trained males; but it is difficult to know how much of this relates to biological differences or to training. At present very few women have attempted to develop their upper bodies to the extent that males do. Measures of endurance reveal no significant differences between highly trained women and men long distance runners when comparisons are made relative to lean body weight.

The anabolic steroids used by male athletes do increase muscle bulk and body weight, and this would be expected to help the athlete who depends on size and brute strength. Scientific studies of strength changes in athletes who use these steroids have produced inconsistent findings. Some show increases in strength in comparisons with control groups, but others do not. Male athletes often take 5 to 10 times the recommended daily dosage of these steroids; hence the effects should be more prominent than the effects of testosterone in the normal male. (Men who use anabolic steroids run the risk of liver damage and possible tissue damage in the testes and prostate.)

At present, it is impossible to predict how close female athletes will come to closing the performance gap that has traditionally separated the sexes. It is clear that many of the current physical differences between females and males result from cultural and societal restrictions that channel most females into sedentary roles after puberty. When males adopt sedentary life-styles in their thirties, they too show a similar loss of strength and endurance.

at college but have regular cycles when they are home for the summer.

Sometimes an unmarried woman who has intercourse and fears pregnancy will have a "late" menstrual period. This delay seems related to her emotional state, but we cannot usually rule out the possibility that she has indeed conceived but has spontaneously aborted very early.

### Proliferation

The menstrual cycle can be divided into four phases: the proliferative phase, ovulation, the secretory phase, and menstruation. The phases of the menstrual cycle, like the events of puberty, are essentially under the control of hormones. The proliferative phase begins when menstruation ceases and lasts for about 9 or 10 days in a 28-day cycle (*see* Figure 4.4). During this time the endometrium that was shed during the preceding menstruation grows and proliferates. The proliferative phase is sometimes called the *preovulatory phase* signifying that it occurs before the release of a mature egg from the ovary. During menstruation and proliferation the pituitary gland secretes FSH, which stimulates several of the ovarian follicles to mature (*see* Figure 4.5) and stimulates these follicles to produce estrogen. The estrogen is

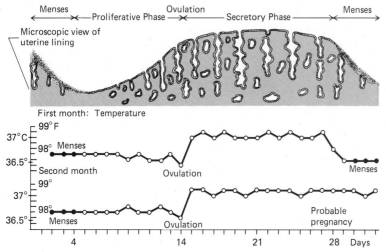

Menses ←——×——Proliferative Phase —→×← Ovulation —→×—— Secretory Phase ——→×←—— Menses →

**Figure 4.4** The phases of the idealized 28-day-menstrual cycle. The last four days represent the beginning of a new menstrual cycle. Adapted from Benson, *Handbook of Obstetrics and Gynecology,* 3rd ed. (Los Altos, Calif.: Lange Medical Publications, 1968), p. 27. Reprinted by permission.

responsible for the regrowth of the endometrial lining. All but one of the follicles decompose and are reabsorbed by the body. The one remaining follicle will ultimately rupture during ovulation in response to the stimulus of increased production of LH by the pituitary gland.

Estrogen stimulates the growth of the glandular surface of the endometrium, and a thickness of about ⅛ inch (3.5 millimeters) is attained during this phase. In addition, estrogen stimulates the size and productivity of the mucous glands of the cervix. Cervical mucus produced under estrogen stimulation is plentiful, thin, and highly viscous; it has an alkaline pH and contains nutrients that can be used by the sperm. As mentioned previously, rising estrogen levels also stimulate growth of the vaginal lining so that maximum thickness is achieved at ovulation.

### Ovulation

The amount of fluid within the maturing follicle increases steadily during the proliferative phase, and ultimately the follicle ruptures and the mature egg (ovum) is released into the fallopian tube. This process is *ovulation,* which occurs about 14 days before menstruation. In a 28-day cycle ovulation thus occurs on about the fourteenth day; but

in a 34-day cycle it occurs on about the twentieth day. The significance of these schedules to those depending on the "rhythm method"[12] for birth control is obvious.

### Secretion

The period after ovulation is called the *secretory,* or *postovulatory,* phase of the menstrual cycle. The pituitary gland, responding to the stimulus of increased estrogen levels in the bloodstream, produces more LH, which travels through the bloodstream to the recently ruptured ovarian follicle and stimulates the remaining cells of the follicle to develop into a microscopic glandular structure called the *corpus luteum.* The corpus luteum, under continued FSH and LH stimulation, produces progesterone, along with estrogen. This new source of estrogen accounts for the second peak in the estrogen level seen in Figure 4.5. (Figures 4.6 shows FSH and LH levels.) Under the combined influence of estrogen and progesterone, the glands of the endometrium that have developed during the proliferative phase become functional and *secrete* nutrient fluid by the eighteenth day of a 28-day cycle, corre-

---

[12]Abstinence from intercourse during the presumed fertile period of the menstrual cycle.

sponding to the time when the ovum is free within the uterine cavity and dependent on uterine secretions for nourishment. In addition, there is increased proliferation of blood vessels within the uterus during this phase.

Progesterone inhibits the flow of cervical mucus that occurs during ovulation and diminishes the thickness of the vaginal lining.

If fertilization does not occur, the pituitary gland responds to the increased blood levels of estrogen and progesterone much as a thermostat responds to an increase in temperature and shuts off production of FSH and LH, depriving the corpus luteum of the chemical stimulation to produce estrogen or progesterone. The corpus luteum then withers away.

### Menstruation

The final phase of the menstrual cycle is actual menstruation, or *menses,* the shedding of the endometrium through the cervix and vagina. Menstruation represents the results of the hormonal changes that happened during the last menstrual cycle. It also signifies the beginning of the next menstrual cycle. The trigger for menstruation appears to be the fall in the estrogen level at the end of the cycle. (The change in progesterone level that

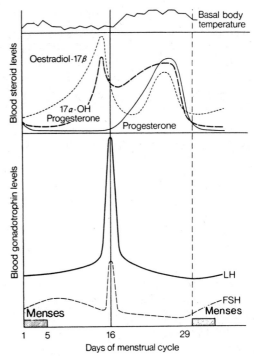

**Figure 4.6**  Hormone changes in the human menstrual cycle.

occurs simultaneously is not necessary for menstruation.)

The menstrual discharge consists primarily of blood, mucus, and fragments of en-

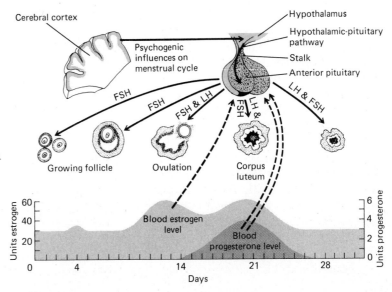

**Figure 4.5**  Ovulation during the menstrual cycle. The long feedback loops are shown by broken lines. Adapted from Benson, *Handbook of Obstetrics and Gynecology,* 3rd ed. (Los Altos, Calif.: Lange Medical Publications, 1968), p. 26. Reprinted by permission.

dometrial tissue. About 2 ounces (60 milliliters) of blood are lost in an average menstrual period. For a woman on an adequate diet, such blood loss can be easily tolerated. Heavy menstrual bleeding in a woman on a deficient diet can, however, lead to anemia. The duration of menstrual bleeding is usually from three to seven days, and periods longer than one week usually indicate some abnormal condition.

Traditionally, women have used externally worn sanitary napkins to absorb the menstrual discharge, but the trend now is toward internally worn tampons—cylinders of absorbent cotton that are inserted into the vagina and removed by an attached string. When tampons were first introduced in 1933, the public and some church leaders became very concerned about possible abuses of the tampon and claimed for some years that it was an instrument of contraception, masturbation, or defloration.[13] It is now realized that these claims are all untrue, and tampons are becoming increasingly popular. Tampons can be used whether or not a woman has had intercourse, since most hymens will easily allow their insertion.

As the estrogen level continues to fall, the pituitary gland again responds as a thermostat would and turns on its secretion of FSH, which increases ovarian production of estrogen that is needed for proliferation of the endometrial wall to begin again. The timing of menstrual cycles, however, is not determined solely by internal hormonal events (see Box 4.3).

## Menstrual Problems

There are two common problems related to menstruation: premenstrual tension and painful menstruation. The degree to which these problems are caused by biological or psychological factors has not been determined. Probably both biological and psychological factors are complexly interwoven, affecting

[13]Delaney et al. (1976).

each woman to different degrees. At present the relative importance of the two factors for each woman cannot be known.

### Biological Effects

Painful menstruation (*dysmenorrhea*) is the most common problem related to the menstrual cycle. Considering the number of changes that a woman's body goes through each month, it is not surprising that problems sometimes accompany these changes. Almost all women at one time or another experience discomfort in the pelvic area during a menstrual period, and for some women the pain can be frequent and severe. Some women have no symptoms, others continue their normal activities with little more than the help of aspirin, and a few may be forced to stay in bed for a day or two during each menstrual period. The basic complaint of women with this condition is cramps in the pelvic area, but there may also be headaches, backaches, nausea, and general discomfort. In women for whom dysmenorrhea is a monthly problem, it usually begins at an early age; it is relatively uncommon for a woman to have her first painful menstrual period after the age of 20. Those women who do suffer from menstrual cramps often notice a great improvement after childbirth, but the reason for this "cure" is unknown.

The cause of menstrual cramps seems to be related to uterine spasms. It seems that these spasms are caused by chemicals in the menstrual fluid known as *prostaglandins* (*see* Chapter 6).[14] Masters and Johnson reported that some women also use orgasm as a means of helping relieve dysmenorrhea. The

[14]Some women's health care centers use "menstrual extraction" for women with painful or difficult periods. The menstrual fluid and tissue are removed shortly after the onset of the menstrual flow by means of a flexible plastic tube inserted via the cervix. Suction is provided by a syringe attached to the end of the tubing. This so-called "five-minute period" carries with it certain hazards and is contraindicated in women with a history of uterine infections, tumors, and a number of other conditions. The medical profession tends to take a dim view of this procedure as practiced by self-help groups.

## Box 4.3   Vaginal Odors as Reproductive Signals

The vaginal odors of some animals play an important role in communicating to the male information about the female's level of sexual receptivity. The female hormone cycle determines which odors are present at which time. Recent studies have shown that women secrete chemical signals—in the form of volatile fatty acids—similar to those of other primates.[1] As shown in the illustration, there is a peak output of the volatile fatty acids shortly before ovulation—except in pill users, who have no peak. People who are allowed to smell odors from various parts of the menstrual cycle can tell the difference between the odors from various points in the cycle. The peak odors are cheeselike due to butanoic acid. When men and women evaluate the vaginal odors, they tend to find them most pleasant during the ovulatory period.[2]

Women are clearly influenced by the odor cycles of other women, although the influence does not operate at a conscious level. Women who live together in college dormitories or other close quarters often begin to menstruate close to the same time.[3] Odor cues are implicated. When a "donor" woman wears cotton pads under her arms for 24-hour periods and other women are exposed to the odors, the menstrual periods of the recipient women shift to become closer to the periods of the donor.[4] Recipients who began an experiment with menstrual periods 9.3 days away from the donor gradually shifted to be only 3.4 days away from the donor after 4 months of odor exposure. It appears that odors from many parts of the body are influenced by the monthly hormonal cycles and can serve to synchronize other women.

Why do women's menstrual cycles become synchronized? In the evolutionary past, reproductive synchrony paid off in terms of survival. In many primate groups, females bear their offspring in one restricted period of the year, when food abundance is optimal for the survival of the young. By having all the females in a group bear infants at the same time of year, all the females are lactating and attentive to infant care at the same time. Because none of the females is sexually receptive, the disruptive influences created by intrusive males and copulatory activity are minimal.[5] Mechanisms that allow females to respond to environmental cues and to each other increase the precision of reproductive synchrony.

Are men influenced by women's odor cycles? Males of many mammalian species are most attracted to females who transmit odors associated with ovulatory and pre-ovulatory periods.[6] The mere presence of sexually cycling females increases spermatogenesis and testosterone production. Although current research has not demonstrated such effects in human males, it is possible that minor effects may still be discovered.

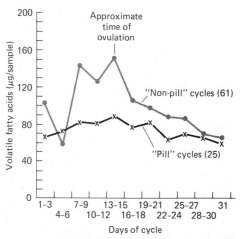

Volatile fatty acids in the vaginal secretions of 47 women (86 cycles). Redrawn from Michael et al. (1974).

[1]Bonsall and Michael (1978).
[2]Doty et al. (1975).
[3]McClintock (1971).
[4]Russell et al. (1977).
[5]Rowell (1972), p. 115.
[6]Michael et al. (1976).

contractions of orgasm increase the rate of menstrual flow, and some women have reported that orgasm experienced shortly after the onset of menstruation not only increases the rate of flow but reduces pelvic cramping when present and frequently relieves menstrually associated backaches.

In addition to pelvic pains, a few women experience so-called "menstrual migraines" with their periods. These headaches are essentially the same as are other migraine headaches, except that they occur particularly during menstruation or just before the beginning of menstrual flow.

During the four to seven days preceding menstruation most women experience symptoms that make them aware of their coming period. When these symptoms are particularly severe they are known as *premenstrual tension,* or the *premenstrual syndrome.* Fatigue, irritability, headache, pain in the lower back, sensations of heaviness in the pelvic region, and a weight gain of as much as several pounds (due to retention of fluids) are among the symptoms. For some women these problems can contribute to emotional and psychological upset at this time. Women who commit crimes or suicide do so more often in the week just before their periods or during their periods. (This is not to say that women are prone to commit crimes or suicide, however, since males commit more than three-fourths of serious crimes and their suicide rate is three times higher than the female rate.) In France a woman who commits a crime during her premenstrual period may use this fact in her defense, claiming "temporary impairment of sanity." The symptoms of premenstrual tension are probably related to the shift in hormone levels occurring at this time, but other factors are also involved.

A separate condition also related to cyclical changes in hormones and retention of fluid is premenstrual pain and swelling of the breasts, known as *mastalgia.* This condition is less common than, but may occur in conjunction with, premenstrual tension.

An unexplained phenomenon is *Mittelschmerz,* pain occurring in midcycle during ovulation. This symptom is most common in young women. It consists of intermittent cramping pains on one or both sides of the lower abdomen lasting for about a day. Occasionally the symptoms of ovulation have been mistaken for appendicitis, much to the embarrassment of the surgeon who decides to operate and finds a healthy appendix.

## Psychological Effects

Throughout history the causes of menstruation have not been understood. As a consequence, beliefs about this normal bodily process have become intermingled with cultural myths as various peoples have tried to explain and cope with a process that seemed to them to be a strange and even mystical or evil occurrence. In most cultures the menstruating woman has been restricted in her activities, and in some societies she has been completely segregated from other people in order to protect them, especially the men, from the harmful influence of menstrual bleeding.

The writings of Pliny, a Roman historian, reflect many prescientific beliefs about menstruation.

Contact with it turns new wine sour, crops touched by it become barren, grafts die, seeds in gardens are dried up, the fruit of trees falls off . . . the edge of steel and the gleam of ivory are dulled, hives of bees die, even bronze and iron are at once seized by rust, and a horrible smell fills the air; to taste it drives dogs mad and infects their bites with an incurable poison.[15]

Today many societies still consider women to be "unclean" during menstruation. Strict menstrual taboos are commonly practiced among some groups in India.[16] These people believe that a man can become polluted simply by touching a menstruating

[15]Pliny (77 A.D. [1961]).
[16]Ullrich (1977).

woman. A priest is needed to purify him. A woman must warn the man of her condition if he might come closer than three feet of her. If a menstruating woman touches certain substances such as wood, rope, or cloth, the pollution is passed to another person who touches the object at the same time. Thus people cannot cross a bridge until a menstruating woman is off the bridge. Other societies impose different restrictions on women during their periods. The women may have to follow special diets and they may be forbidden to participate in specified activities.

The Old Testament and the Koran and Hindu law books all specifically prohibit intercourse during menstruation because women are presumed to be polluted and dangerous at this time and will defile anyone who has contact with the menstrual discharge. Orthodox Jewish women are not permitted to have sex while menstruating and for seven days afterwards. They must take a ritual bath before they are considered to be "clean" again. Several South American and African tribes believe that a man would become physically ill if he had sexual relations with a menstruating partner. Some nomads of northeast Asia believe that a woman would become sick and eventually sterile if she copulated while bleeding. Other tribes believe that a menstruating woman who had any physical contact with a man could cause him to lose his virility and his hunting ability. In such societies women are physically isolated from their communities during menstruation.

There is no medical reason why a woman should refrain from intercourse while she is menstruating. In fact, the taboos against intercourse during menstruation are declining. A recent survey of 575 married men and women revealed that even though 71 percent of people 55 years old and over had never had intercourse during menstruation, only 28 percent of the people 35 and younger could make the same statement.[17]

A menstruating woman is often thought of as having a sickness or a handicap. In fact, many menstruating women are not completely comfortable with their bodily processes. Most women would feel quite embarrassed if some menstrual blood were to "soil" their outer clothing. If the source of blood were a cut they would be much less embarrassed. Women learn from their culture that menstruation is something about which to feel uncomfortable.

A study of 298 college women found differences in menstrual problems in women of different religious groups.[18] Catholic women often felt that menstrual difficulties were part of their role. The Catholic women (especially virgins) who thought a woman belonged in the home and who had no career desires were most likely to have severe menstrual distress. Among the Jewish women, those who followed the Orthodox Jewish teachings regarding menstruation were most likely to have menstrual difficulties. The Protestant group studied was too diverse for any correlations to be evident. Another finding of the study was that women who reported physical pain and psychological stress during menstruation tended to report more problems at other times, too.

A recent study suggests that to some degree psychological factors are operating in the reporting of premenstrual symptoms.[19] Forty-four college women who were approximately one week from menstruation were led to believe that an electroencephalogram (EEG) could predict the timing of their next menstrual period. After undergoing an elaborate "test" that was supposed to predict the onset of the next period, all the women filled out a Menstrual Distress Questionnaire, the standard questionnaire used in menstrual research. The women who were told that their period was due in one or two days were much more likely to report the following

---

[17]Paige (1978).

[18]Paige (1973).
[19]Ruble (1977).

problems—water retention, pain, and changed eating habits—than women who were told that their period would be ten days away. (A follow-up on the women found that the average number of days before the onset of the next period did not differ between the two groups.) Studies such as these indicate that to some extent there is a psychological component to menstrual problems.

## Hormonal Control of Sexual Behavior

The degree to which sexual behavior in monkeys and apes is under hormonal control is not entirely clear, but the relation of sexual behavior to sex hormones in lower mammals is well known. Females come into estrus—that is, become receptive to the sexual advances of males—at specific points in their cycles that coincide with ovulation and maximum levels of female sex hormones. Mating at this time obviously enhances the probability of reproduction, and there is essentially no sexual activity in lower mammals except when the females are fertile and likely to conceive. Removal of the ovaries of a guinea pig or hamster has been shown to eliminate sexual activity, but the female can be made sexually receptive again with injections of estrogen and progesterone.

The human female is unique in that her sexuality is not limited to one particular part of the menstrual cycle; that is, human sexual activity is not linked to specific periods of the female's cycle, as it is in most other mammals. Studies have shown, however, that there are predictable variations in the frequency of intercourse and orgasm during the menstrual cycle in many women, with peaks occurring at midcycle and just before menstruation.[20] A similar pattern for self-rating of sexual arousal in young women during the menstrual cycle has also been found.[21] Other

[20]Udry and Morris (1968).
[21]Moos, Lunde, et al. (1969).

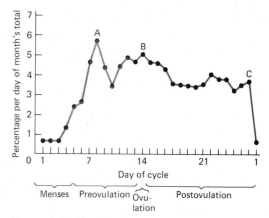

**Figure 4.7** The distribution of intercourse for 52 women on 1123 occasions during the menstrual cycle. Redrawn from McCance et al. (1952).

studies indicate that the frequency of intercourse is highest after menstruation ceases, with minor peaks occurring around the time of ovulation and perhaps right before menstruation (see Figure 4.7).[22] The pattern in which coitus peaks after menstruation and declines thereafter may reflect a rebound of sexual activity after abstinence during menstruation. Interestingly a study that measured female-initiated coitus did not find a large peak occurring after menstruation in those women who engaged in intercourse during their menstrual period. The minor peak before the menstrual period may reflect preparation for the time when the likelihood for intercourse is reduced.

It is possible that changes in coital frequencies may be influenced by variations in sex hormones. The increased frequency of intercourse may be related to the increased levels of estrogen, although it may be related to androgens, which some data indicate may also peak at this time. There is evidence that the androgens secreted by the adrenal glands influence the sex drive of women: Surgical removal of the adrenal glands has a more

[22]McCance et al. (1952), James (1971), Gold and Adams (1978).

predictable effect on sexual interest than does removal of the ovaries. Women who receive androgenic hormones for medical reasons sometimes also report dramatic increases in sexual desire while under treatment. Three possible mechanisms have been suggested to explain this result:

1. Androgens cause heightened susceptibility to psychic and somatic sexual stimulation.
2. Androgens produce increased sensitivity of the external genitalia.
3. Androgens induce greater intensity of sexual gratification (orgasm).[23]

## Menopausal Problems

There are several common problems that women experience during menopause. As with menstruation, these symptoms are probably caused by a complex mixture of biological and psychological factors. At present, the relative influence of the various factors is not known.

### Biological Effects

The term *menopause* refers to the permanent cessation of menstruation due to the physiological changes associated with aging. A broader term, encompassing the various changes that occur in connection with the altered functioning of the ovaries is *climacteric* ("critical period"). The latter generally includes the years from 45 to 60, the years of the "change of life."

Menstruation most commonly ceases between the ages of 46 and 50. Menstrual periods usually become very irregular several years before menopause, and this interval is one of relative infertility, as is that just after menarche. Pregnancy beyond age 47 is rare, but it has been medically documented as late as 61.

There is evidence that menopause comes at an earlier age in women who smoke than

in nonsmokers. In a study of 3500 women in seven countries researchers found that at ages 48 to 49, a woman who smokes a pack a day or more is nearly twice as likely to be past menopause as is a woman who never smoked. The researchers suggest that nicotine may have an effect on the secretion of the hormones involved in menopause or on the way the body handles sex hormones.

The mechanism of menopause, unlike that of puberty, is not related to the pituitary gland. The latter continues to pour out FSH, but for some reason the ovaries gradually fail to respond and very little estrogen is produced.

The best known symptom of the climacteric is the *hot flash* or *flush,* caused by the dilation of blood vessels resulting in the experience of waves of heat spreading over the face and the upper half of the body. It may be followed by perspiration or chills and may last for a few seconds or longer. Other physical symptoms include headaches, dizziness, heart palpitations, and pains in the joints. Severe depression (*involutional melancholia*) may occur at this time in women with no previous history of mental illness. The cause is unknown. Less severe depression during menopause may be partly related to the hormonal imbalance of this period.

Some of the symptoms of the climacteric will affect about three-fourths of women to some degree, but only about 10 percent of them are obviously inconvenienced by these problems. Doctors have prescribed estrogens to relieve these symptoms, but in 1976 the U.S. Food and Drug Administration warned that women who take female hormones to relieve menopause symptoms run a "marked increase" in the risk of cancer of the uterus. The agency cautioned that women should be given the lowest possible doses for the shortest possible time.

Certain changes in the genitals occur during the years after menopause, including a gradual shrinking of the uterus and vaginal

[23]Greenblatt et al. (1972).

lining. Some women experience a new awakening of sexual desire after menopause, perhaps because they no longer need worry about pregnancy or contraception.

### Psychological Effects

Historically, menopause has been a little-understood phenomenon. In some cultures women are given special privileges and status when they cease menstruating and bearing children. In other cultures menopause has been stereotyped as a difficult and trying time. Indeed, women in *Inis Beag,* a rural community in Ireland (*see* Chapter 13) believe that menopause can produce madness, and many withdraw from life when they reach that age. Some even spend the rest of their lives in bed. A century ago in the United States and England some doctors treated women with menopausal complaints by bleeding them or by applying leeches in order to rid them of excess blood. Clearly, if people believe that menopause is a difficult life transition, the belief can become a self-fulfilling prophecy.

It is difficult to determine exactly what role psychological factors play in causing or exacerbating the problems women experience during the climacteric.[24] Depression in particular appears to be influenced by both the hormonal changes and psychological difficulties associated with this period of life. Part of these difficulties can be traced to the problems some women experience in coping with their loss of reproductive capacity, their changing role within the family and society, and their own aging. Although some women have problems when children leave home, several studies have found that middle-aged women whose children no longer live at home report being happier than women of about the same age who still have a child at home.

[24]For a comprehensive review of the biological and psychological causes of menopause see Tucker (1977).

It is clear that menopausal problems result from more than hormonal causes alone. Several studies report no correlation between emotional problems and estrogen levels (as judged from vaginal smears). Estrogen therapy is not an automatic cure for menopausal depression. At present the role of estrogen therapy in relieving depression is controversial. Some studies report very positive results and other studies give more mixed findings. In addition, research has shown that certain social and psychological factors play a key role in determining whether a woman will experience problems during the climacteric. For example, women who overspecialized in the motherhood role experience more problems at menopause than other women. Women who have not been pregnant report fewer problems than women who have. Women from higher economic groups report fewer problems than women from lower economic groups. Better educated women report fewer problems than women with a grade-school education.

In general, it must be emphasized that most women cope successfully with the biological and psychological problems they experience during menopause. The stereotype that menopause is a very difficult period for most women is not true.

## Male Reproductive Endocrinology

### Puberty

Puberty begins somewhat later and lasts longer in boys than it does in girls. It is initiated at age 10 or 11 in boys by the same pituitary hormones that mediate it in girls, FSH and LH. But in males, as was noted earlier, LH is called "interstitial-cell-stimulating hormone" (ICSH) because of the difference in the site of its action.

ICSH reaches the interstitial cells of the testes through the bloodstream and stimulates them, initiating the process of puberty. Little is seen in the way of striking external changes, but the interstitial cells begin producing the primary androgen (male sex hormone), a compound known as *testosterone.* This one substance is essentially responsible for the development of all the physical changes (including development of *secondary sex characteristics*) that occur during puberty (*see* Figure 4.3).

### External Changes

The first noticeable changes resulting from testosterone stimulation include enlargement of the testes and penis and the appearance of fine straight hair at the base of the penis. At first the penis increases in circumference more than in length. By age 12 it is still, on the average, about 1.5 inches (3.8 centimeters) long in the relaxed state and less than 3 inches (7.6 centimeters) long in erection.

As the testes enlarge, however, their capacity to produce testosterone increases, and at age 13 or 14 rapid growth of the penis, testes, and pubic hair begins. By this age most girls will have had their first menstrual period,

although they will not necessarily be ovulating as yet.

Axillary hair does not appear until about age 15 in most boys, and at the same time some fuzz appears on their upper lips. Adult beards do not appear for two or three more years, however, and indeed by age 17 about 50 percent of boys in the United States have not yet shaved and many others shave only infrequently. Continued development of facial and chest hair under androgen stimulation continues beyond age 20 in many young men, along with recession of the hairline (*see* Figure 4.8).

A very noticeable change that occurs during puberty—though strictly speaking an internal change—is the deepening of the voice, related to growth of the *larynx* ("voice box") in response to hormonal stimulation. The deepening of the voice may be gradual or fairly abrupt, but on the average it occurs by age 14 or 15.

Acne becomes a source of embarrassment to many boys at age 15 or 16, as does another transient phenomenon, enlargement

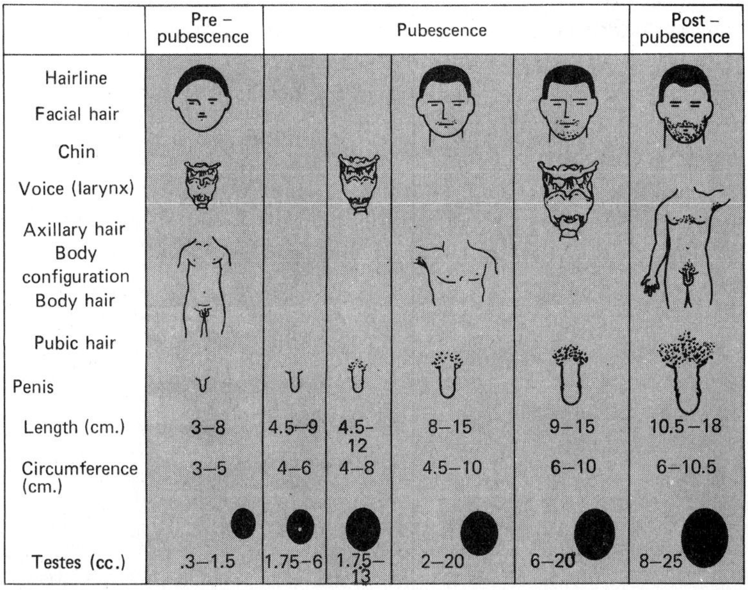

**Figure 4.8** Development of some secondary sex characteristics in men. From Wilkins, Blizzard, and Migeon, *The Diagnosis and Treatment of Endocrine Disorders in Childhood and Adolescence,* 3rd ed. (Springfield, Ill.: Charles C Thomas, 1965), p. 200. Reprinted by permission.

|  | Pre-pubescence | Pubescence | | | | Post-pubescence |
|---|---|---|---|---|---|---|
| Hairline<br>Facial hair<br>Chin | | | | | | |
| Voice (larynx) | | | | | | |
| Axillary hair<br>Body configuration<br>Body hair | | | | | | |
| Pubic hair | | | | | | |
| Penis | | | | | | |
| Length (cm.) | 3–8 | 4.5–9 | 4.5–12 | 8–15 | 9–15 | 10.5–18 |
| Circumference (cm.) | 3–5 | 4–6 | 4–8 | 4.5–10 | 6–10 | 6–10.5 |
| Testes (cc.) | .3–1.5 | 1.75–6 | 1.75–13 | 2–20 | 6–20 | 8–25 |

---

### Box 4.4    Premature Puberty

Although puberty normally begins at about age 12 and 13 for girls and boys, it can begin much earlier. If the changes begin before 8 in girls and 9 in boys, the condition is called precocious puberty.[1] Girls are twice as likely to have precocious puberty as boys. There are cases of menstruation beginning in the first year of life. The youngest known mother was a Peruvian girl who began menstruating at 3 years and gave birth (by cesarean section) at the age of 5 years and 7 months. There are cases in which penile development began at 5 months and spermatogenesis at 5 years. Boys as young as 7 years of age have reportedly fathered children. In many cases, especially in girls, precocious puberty does not indicate any major problem: The early puberty merely reflects a natural variation in the body's timing mechanisms. However, 20 percent of girls and 60 percent of boys with precocious puberty have a serious organic disease. Tumors in the hypothalamic region of the brain, the gonads, or the adrenals can trigger the production of gonadotrophins or sex hormones, which in turn lead to the early sexual maturation of the body.

In some cases, a child may undergo incomplete precocious puberty. There may be early breast development or early growth of pubic hair, but no other changes. These changes are less likely to be related to major organic disorders than is complete precocious puberty.

[1]Katchadourian (1977), pp. 128 ff.

---

of the breasts, or *gynecomastia*. The latter occurs in about 80 percent of pubertal boys and is probably related to small amounts of female sex hormones produced by the testes.

Overall growth in both height and weight occurs during puberty. But, whereas the girl develops fat deposits in breasts and hips at this time, the boy's new weight is in the form of increased muscle mass and he consequently develops a quite different physique.[25] Whereas the female pelvis undergoes enlargement at puberty, more striking expansion occurs in the boy's shoulders and rib cage at this time. There is also a definite spurt in the linear growth rates of boys at puberty, and this process continues until age 20 or 21, when the male hormones finally put the brakes on growth in the long bones of the skeleton. As previously noted, female sex hormones induce this "braking" process more rapidly, which accounts for the generally smaller height of women.[26]

### Internal Changes

Whereas the changes in external appearance just described give the appearance of biological maturity to the growing boy, significant internal changes must also occur before the capacity for reproductive activity is actually achieved.

The pituitary hormone FSH, which stimulates maturation of the ovum in the female, is also essential in the production of mature sperm in the male. With the increase in FSH secretion during puberty, germ cells in the lining of the seminiferous tubules of the testes (*see* Chapter 2) begin to divide and

---

[25]The use of supplemental androgens ("anabolic steroids") to increase the muscle mass of Olympic athletes (particularly shotputters) and professional athletes (particularly football players) has become widespread and controversial in recent years.

[26]This statement does not imply that a tall girl has a deficiency of female sex hormones or that a tall boy has a deficiency of male hormones. Other factors, including heredity and nutrition, influence stature. Americans of both sexes have been getting progressively taller with each successive generation during the last century, a phenomenon usually ascribed to a better diet.

**Box 4.4**   continued

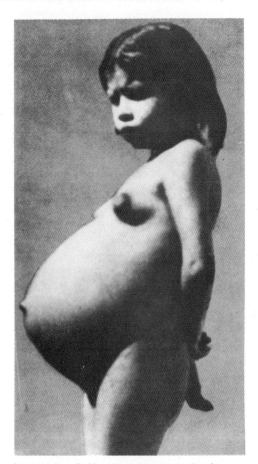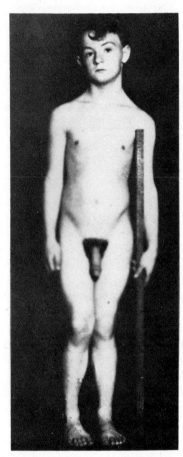

Lina Medina (left) is the youngest mother known in the medical literature. She gave birth in 1939 at the age of 5 years and 7 months. A 5-year-old boy with an unusual adrenal development (right) has the height of an 11.5-year-old and precocious penile development. From Wilkins, Blizzard, and Migeon *The Diagnosis and Treatment of Endocrine Disorders in Childhood and Adolescence*, 3rd ed. (Springfield, Ill.: Charles C Thomas, 1965). Reprinted by permission.

differentiate into mature sperm. Mature sperm are first present in the ejaculatory fluid at age 15, on the average, but, as with the other changes at puberty, there is wide variation among individuals, encompassing the range from ages 11 to 17.

FSH is a necessary but not sufficient stimulus for the production of mature sperm. Other hormones, particularly thyroid hor-

mone, must also be present in sufficient concentrations. In addition, a temperature lower than normal body temperature is essential to mature sperm production. In undescended testicles, which are still in the abdominal cavity, mature sperm cannot be produced because of the warmth of the environment.

The organs that supply fluids for the semen enlarge significantly during puberty. The prostate gland is particularly sensitive to the stimulation of testosterone. By age 13 or 14 the prostate gland is producing fluid that can be ejaculated during orgasm, but this fluid will probably not contain mature sperm. Just as the menarche does not necessarily indicate that a girl has ovulated, so the ability to ejaculate does not indicate that a boy is fertile.

At about this same age a boy begins to have "wet dreams," or *nocturnal emissions* of seminal fluid (*see* Chapter 9).

### Becoming an Adult

Many societies have specific puberty rites for boys, generally to signify that they have reached adulthood. As there is no event for boys comparable to menarche, an arbitrary age like 13 may be taken as the time for initiation. In some societies, however, particular manifestations like the appearance of facial hair or the first "wet dream" are required for adult status. Women are often excluded from such male puberty rites, and sexual information is often transmitted to boys during these ceremonies. In addition, circumcision may be performed at this time (*see* Chapter 2). Many tribal peoples require the performance of certain acts of physical strength, skill, and endurance; in addition, a boy may be required to participate in certain sexual activities.

With the exception of the bar mitzvah among Jews, our society does not have formal puberty rites for boys. Nevertheless, like girls, boys are very sensitive to the physical transformations that occur in their peers during puberty and respond accordingly. Early-maturing boys have thus been found to have greater peer-group prestige and are most often ranked as "leaders" among boys of pubertal age. It should be noted, however, that in adulthood the late maturers are often found to be more adventurous, flexible, and assertive than are early maturers.[27]

[27]Hamburg and Lunde (1966), p. 4.

## Hormones and Fertility in the Male

There is no fertility cycle in men as there is in women. Sperm are normally produced throughout the reproductive years, and the healthy man is consistently fertile at all times of the month and year. This phenomenon corresponds to the fairly consistent secretion of testosterone by the testes, in contrast to the cyclical secretion of sex hormones in the female.

In the animal world there are some exceptions to this continuous male fertility. Some species, like deer and sheep, have specific "rutting seasons" during which the males are fertile and sexually active. For the rest of the year these animals are infertile and generally uninterested in sexual activity. In certain wild rodents the testes are actually drawn into the body cavity except during the mating season, when they descend into the scrotum. During the period that the testes are inside the abdominal cavity they produce neither sperm nor male sex hormones.

Although testosterone secretion and spermatogenesis are generally constant in men, variations may occur in certain situations. Severe emotional shock has been known to cause temporary cessation of sperm production in some men. Studies of soldiers in combat show that testosterone levels rise and fall in response to the degree of stress to which men are exposed.[28] Recent studies have also shown that intercourse in males is followed by an increase in the level of testosterone in the bloodstream.[29] These findings should not seem surprising in light of the interrelations of sex hormones, pituitary hormones, and the brain, which we described earlier in this chapter.

Other factors may also affect fertility in men, either temporarily or permanently. Certain drugs, including some used in the treatment of cancer and at least one antibiotic

[28]Rose (1969), p. 136.
[29]Fox et al. (1972).

**Box 4.5**    Eunuchs

Castration, the removal of the testes, can occur before puberty or after, by accident or by plan. Although castration is rare in modern times (except as a result of accidents, disease, or war), it used to be a much more common practice.

Eunuchs who are castrated before puberty retain their high voices and do not develop the secondary sex traits of beards and body hair.[1] They develop subcutaneous fat deposits that follow the feminine pattern. Men who were castrated before puberty can have erections, much the same as prepubertal boys can. Once prepubertal castrates reach adulthood some of them may marry and have regular sexual intercourse and satisfying erotic feelings.[2]

Early castration was most common in the Near East, and probably first performed in ancient Egypt where hundreds of young boys would be castrated in a single religious ceremony and their genitals offered as sacrifices to the gods. A eunuch (from Greek *eunoukhos*, "guardian of the bed") is a castrated male; and eunuchs have perhaps been most famous as harem guards. Although most eunuchs were slaves to Muslim rulers, they occasionally worked themselves into positions of great influence and power. The sex drive varied greatly among eunuchs, but some were apparently very active sexually.

The practice of using castrates in choirs spread to the West from Constantinople.

*Castrati* were used in the papal choir until the early nineteenth century. Because women were not allowed to perform in the opera up into the eighteenth century in parts of Europe, talented boys were castrated to produce the sopranos who would eventually play women's roles.

Eunuchs played a powerful role in Chinese imperial governments for more than 2000 years and through 25 dynasties.[3] At the peak of their influence, there were over 70,000 eunuchs in the service of the Ming dynasty. They gained great power and wealth by manipulating the emperors and were in part responsible for the decline and fall of every dynasty. The eunuch system was not completely abolished until 1924, when the final 470 eunuchs were driven from the last emperor's palace.

Men who are castrated after puberty do not lose the secondary sex traits they gained during puberty, nor do they lose the sexual knowledge and psychosexual orientation they developed in prior years. There is a great deal of variability among these men in their sexual interest and performance. The ability to ejaculate is lost, since it is hormonally dependent; but orgasm without ejaculation feels much the same as the normal orgasm. Most men lose their sexual potency, but some do not.

[1]Money and Ehrhardt (1972).
[2]Money and Alexander (1967).
[3]Mitamura (1970).

(*Furadantin*), may cause temporary sterility. A prolonged exposure to high temperature, as in an illness involving a high fever or in frequent and prolonged hot baths, may inhibit sperm production. Mumps may cause permanent sterility in men through the mechanism described in Chapter 2. Severe or prolonged radiation can cause sterility or the production of abnormal sperm. In the days when fluoroscopes were commonly used in fitting shoes, some shoe salesmen were rendered sterile by prolonged exposure to radiation.

## Aging in the Male

There is no male equivalent of the menopause in the female. The term *menopause* means the cessation of the menses. Thus, males do not have a menopause in the true

## Box 4.6   Hormonal Errors

Individuals who exhibit external genital characteristics of both sexes have traditionally been called *hermaphrodites*. Ancient Greek and Roman art and literature are replete with images and reference to the deity Hermaphroditus. Herodotus and Plato both referred to an ancient tribe living north of the Black Sea said to be ambisexual. Plato theorized that such people represented a third sex, which had died out over time. To other writers, however, hermaphrodites seemed to have supernatural qualities. From the time of Theophrastus (382–287 B.C.) through the Roman Empire, Hermaphroditus was considered a god, the offspring of Hermes (god of occult science) and Aphrodite (goddess of love).

In medical terminology a true hermaphrodite (*hermaphroditus verus*) is an individual who has both ovarian and testicular tissue. The condition is extremely rare: Only 60 cases have been reported in the entire world medical literature of the twentieth century. Such an individual may have one ovary and one testicle or sex glands that contain mixtures of ovarian and testicular tissues. This combination of tissues and hormones is usually accompanied by a mixture of masculine and feminine characteristics, but hermaphrodites usually have masculine genitals and feminine breasts. There often is some sort of vaginal opening beneath the penis, and many hermaphrodites menstruate. Development of the uterus is often incomplete—with, for instance, only one fallopian tube present. True hermaphrodites are usually genetic females (XX). They can be raised as males or females and the decision is often made based on their appearance at birth. The first picture shows a hermaphrodite who is a genetic female, with a right testis and left ovary. This person has always lived as a male. He is married and a stepfather.

In addition to true hermaphrodites, there are also male and female pseudohermaphrodites. Unlike the true hermaphrodite, the pseudohermaphrodite does not have both ovarian and testicular tissue. But pseudohermaphrodites do have some sex organs that resemble the organs of the opposite sex. The development of both male and female sex organs is caused by the simultaneous presence, in significant amounts, of male and female sex hormones during embryological development (*see* Chapter 2).

Since male sex hormones can also be produced by the adrenal glands, a female may present a masculine appearance because of a genetic defect in the adrenal glands that results in the pouring forth of large amounts of androgens. This condition is known as the adrenogenital syndrome

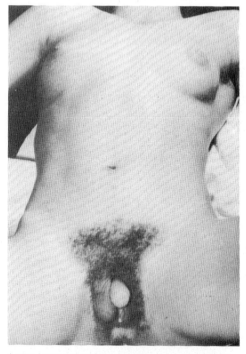

A true hermaphrodite, who is a genetic female that has always lived as a male. He has one ovary and one testis. From Money *Sex Errors of the Body.* (Baltimore: Johns Hopkins Press, © 1968), p. 121. Reprinted by permission.

## Box 4.6 continued

(AGS). Such individuals not only have masculine-appearing external genitals at birth, but they also become further masculinized at puberty due to the continued secretion of large amounts of androgens.

Although such a woman may have normal internal female genitalia, including ovaries, her clitoris is often enlarged and resembles a penis as can be seen in the picture below. In addition, the folds of the labia may be fused in such a manner as to resemble a scrotum, with the result that such a female may be reared as a male.

There are other conditions that can produce a female pseudohermaphrodite. If a mother has an androgen-producing tumor while pregnant or the mother is given progestin (synthetic progesterone, which sometimes has male effects) while pregnant, the fetal genitals may be masculinized, and a female pseudohermaphrodite may be created. Unlike the female with AGS, there is no excess of male hormones present in the female's

body after birth; thus no further masculinization will take place.

Male pseudohermaphrodites are produced during intrauterine development if the male does not produce sufficient male hormones or if these hormones are not biologically active. For example, some male pseudohermaphrodites have androgen insensitivity. In this case a genetic male (XY) with androgen-secreting testes (usually undescended) has the external

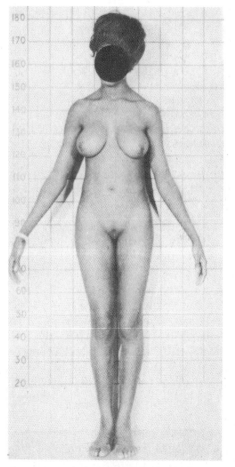

A woman with the androgen insensitivity syndrome. She is a genetic male. From Money *Sex Errors of the Body.* (Baltimore: Johns Hopkins Press, © 1968), p. 113. Reprinted by permission.

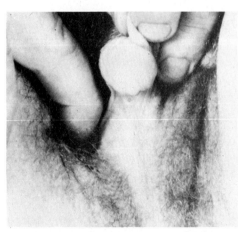

Characteristic external genitalia of a female pseudohermaphrodite. From Jones and Scott, *Hermaphroditism, Genital Anomalies and Related Endocrine Disorders* (Baltimore: Williams & Wilkins, 1958), p. 212. © 1958 by The Williams & Wilkins Co.; reprinted by permission.

**Box 4.6**    continued

appearance of a female, a vagina, but no uterus. It would appear that receptor sites for androgens, located on the X chromosome and/or other chromosomes, are blocked by some unknown mechanism, thereby rendering the male hormones inactive.[1] Such males are sometimes raised as females. Others are raised as males, but they do not develop the secondary sex traits of the male since their bodies do not respond to androgens. Instead their bodies respond to the female hormones present and breasts develop. The third picture is of a female who is actually a genetic male who has the androgen-insensitivity syndrome.

There are many variations among male pseudohermaphrodites, from males with androgen insensitivity to males with almost completely normal appearances. For example, a male may have normal external genitals but also have a uterus. A male may have a clitoris-like penis with the urethra opening at the base. The testes may be undescended and a vaginal opening may be present.

In most cases the gender identity and gender role of hermaphrodites and pseudohermaphrodites corresponds to the sex they were raised to be. Also, the sexual activity of both true and pseudohermaphrodites reflects their assigned gender roles—those reared as males usually choose female sexual partners, and vice versa. (There is one unusual case on record, however, of a hermaphrodite who had intercourse alternately with men and women, using the vaginal opening or the enlarged clitoris to suit the occasion.)

Most types of pseudohermaphroditism can be corrected by surgical or hormonal and surgical treatment; but it is important that the proper diagnosis be made early. Psychological difficulties are quite common in individuals who have lived for several years or more with mistaken or ambiguous sexual identities, as might be expected.

[1]Mittwoch (1973).

sense of the word. Whereas the ovaries essentially cease to function at a fairly specific time in a woman's life, the testes continue to function in men indefinitely, though there may be gradual declines in the rates of testosterone secretion and sperm production (*see* Figure 4.9). There are well-documented reports of men of 90 years of age having fathered children, and viable sperm have been found in the ejaculations of men even older.

Although older men do not technically experience a biological menopause, there are certain problems that become more common in men with age. The most common biological problem is enlargement of the prostate gland, which occurs in 10 percent of men by age 40 and in 50 percent of men by age 80.

This enlargement interferes with control of urination and causes frequent nightly urination, but it can usually be corrected. It also causes men to worry about their aging and sexual ability (*see* Chapter 7).

Men may experience psychological changes in their fifties and sixties.[30] Any of a variety of life experiences that are common at that age can cause some men problems, although other men will not be affected. By their fifties, people are often aware of growing old, and thoughts of death may be painful. It may hurt to see younger men moving into exciting jobs while the older men—past their prime—are being put out to pasture. Many

[30]Rubin (1977).

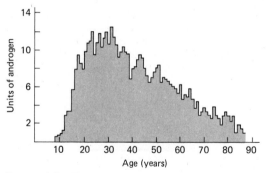

**Figure 4.9**  The quantities of biologically active androgens excreted by boys and men. Redrawn from Pedersen-Bjergaard and Tónnesen (1948).

men who once had high aspirations begin to realize that they had not accomplished as much in life as they had hoped. Changes in the family structure—as the children leave home—may leave the husband and wife alone together, often aware that they have grown apart during the child-rearing years.

Naturally, not all men experience these changes or find aging to be problematic. Numerous factors can help make aging easier to cope with: among these are a good marital and family relationship, health and vigor, and a realistic appraisal of one's abilities and acceptance of that which is inevitable. There is no change related to hormone levels that need interfere with sexual activities in men or women as they grow older. Both men and women can continue to have fulfilling sex lives well into old age.

# Chapter 4

# Conception, Pregnancy, and Childbirth

For centuries the "miracle of birth" must have seemed just that—a miracle. No one could fully explain how it happened, and in some societies people did not even make the connection between sexual intercourse and pregnancy. Two South Pacific tribes, the Arunta and the Trobrianders, believed that intercourse was purely a source of pleasure unrelated to pregnancy. Natives of these tribes believed that conception occurred when a spirit embryo entered the body of a woman. The spirit was believed to enter the uterus either through the vagina or through the head. In the latter instance, it was thought to travel through the bloodstream to the uterus.

Other people, though recognizing that intercourse is essential to conception, believed that supernatural powers determined whether or not specific acts of intercourse would cause pregnancy. The Jivaro tribesmen in the Amazon region believed that women were particularly fertile during the phase of the new moon. Perhaps this belief reflects association of the lunar cycle with the menstrual cycle, both of which are about the same length.

A few tribes, like the Kiwai of New Guinea and the Baiga of India, believed that a woman could become pregnant from something that she had eaten, implying that a substance taken in through the mouth may have contained semen, which then found its way to the uterus.

Today we know that it is the union of sperm and egg that results in conception, pregnancy, and the birth of a child. Even though there are still aspects of these processes that have yet to be fully explained, the miracle of birth is no longer quite as mysterious as it once must have seemed.

## Conception

The moment when sperm and egg unite is the moment of conception. In order to understand exactly what this means, it is necessary to have some basic knowledge of the cells involved—sperm and eggs.

## Sperm

The persistence of myths and misconceptions about conception is understandable when we realize that sperm were not discovered until 300 years ago, and even then there was little agreement on what sperm were. The discovery was made in the laboratory of Anton van Leeuwenhoek (1632–1723), a Dutchman who did some of the first scientific investigations of life forms with a microscope. He noted that sperm resembled other microscopic organisms he had seen swimming

## Box 5.1   Where Do Babies Come From?

People have always been curious about the origins of human life. Early observers often allowed their imaginations to carry their theories beyond the facts.[1] Over 2000 years ago Aristotle speculated that the human fetus originated from a union of menstrual blood and male seminal fluid. After the microscope was developed in the late 1600s, scientists could see the sperm in the seminal fluid for the first time. Naturally, sperm did not resemble humans in appearance. Nor could any structures with human form be found inside the female body. The absence of human form in either the sperm or the ovaries created more questions than answers: How could a human baby develop from formless origins?

Two schools of thought emerged to explain the inexplicable. The ovists claimed that a minuscule, but fully formed, baby was contained in the eggs and that the sperm functioned only to activate the growth of the preformed baby. The homunculists held the opposite view, that the preformed baby resided inside the head of the sperm but did not begin to develop until it arrived in the fertile environment provided by the uterus. Looking through their crude early microscopes, several of the homunculists claimed to have seen a homunculus—or little man—inside the sperm, which served

[1]Meyer (1939).

as evidence for their theory. Large sperm were presumed to produce male children, and small sperm to produce females.

The logic of both the ovist and homunculist theories led to the incredible conclusion that all generations of humans had been preformed from the beginning and that each generation was stacked inside either the ova or sperm of all previous generations, like Chinese boxes!

Of course, these early scientists could not know that fetal development is under control of genetic information. This modern theory was not developed until the early 1900s.

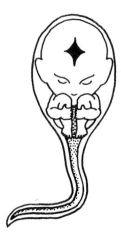

A homunculus as drawn by Niklaas Hartsoeker in 1694.

about in pond water, and he named them *spermatozoa* ("seed animals"). Van Leeuwenhoek was convinced that sperm were involved in human reproduction, but others at the time thought that sperm were only miscellaneous organisms contaminating the seminal fluid. A hundred years later Lazaro Spallanzani (1729–1799) demonstrated that seminal fluid from a male dog that had been filtered (thus removing the sperm) was incapable of causing pregnancy when injected into the vagina of a female dog.

From the union of male and female germ cells one new cell is formed. It divides into two cells, then into four cells, then into eight cells, and so on. If all goes according to plan, the result after approximately nine months is a fully formed human infant. The process of cell division is called *mitosis*. The information that guides cells in their division and multiplication is contained in the *chromosomes*, threadlike bits of material in the nucleus of each cell.

Human body cells contain 46 chromosomes: 22 pairs and two "sex chromosomes" that determine the sex of an individual (*see* Figure 5.1). The body cells of females have two X chromosomes, and those of males have one X and a smaller Y chromosome. Human germ cells (sperm and egg) are different from other cells. Instead of containing the usual 46 chromosomes, they contain only 23. Thus, when sperm meets egg and their chromosomes combine, the result is one cell with the usual 46 chromosomes. Germ cells have only half the normal number of chromosomes because they go through a special type of reduction division called *meiosis* (from the Greek word for "less").

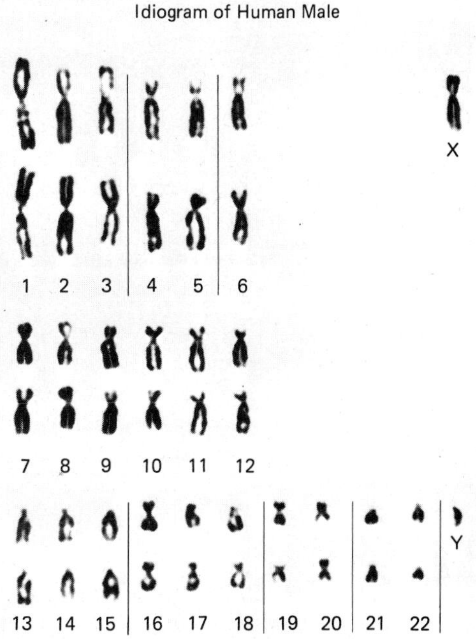

### Sperm Production

Sperm are produced in the seminiferous tubules of the testes, and they go through several stages of development beginning with cells called *spermatogonia* that lie along the internal linings of the tubules. The spermatogonia divide into *spermatocytes*, which in turn undergo reduction division, so that the resulting *spermatids* have only 23 chromosomes. Because this process begins with normal male cells (having both an X and a Y chromosome), half of the spermatids will have an X and half will have a Y chromosome after meiosis. In females there are two X chromosomes in each cell. So, when meiosis occurs during the maturation of eggs, each egg ends up with one X chromosome.

It is the spermatid that eventually becomes a mature sperm. Its pear-shaped head contains the nucleus of the cell. The spermatid has a cone-shaped middlepiece and a tail that enables it to swim (*see* Figure 5.2). The head of the sperm is about 5 microns long, the middlepiece another 5 microns, and

Idiogram of Human Male

**Figure 5.1**  The human chromosomes (autosomes 1–22, sex chromosomes X and Y) from a male cell in culture. The 46 chromosomes are from a photomicrographic print (cut single) arranged in pairs, and grouped according to sizes and relative lengths of the arms.  From Tjio and Puck, *P.N.A.S.* 44 (1958), 1232.

the tail 30 to 50 microns (1 micron equals 0.000039 inch or 0.001 millimeter). Enough sperm to repopulate the world would fit into a space the size of an aspirin tablet.

As these tiny sperm cells are produced they move up the seminiferous tubules to the epididymis (*see* Figure 2.18) and are stored there until they are ejaculated. If ejaculation does not occur within 30 to 60 days, the sperm degenerate and are replaced by the new ones that are being produced continuously.

### The Journey of the Sperm

Billions of sperm are ejaculated by most men during their lifetimes, but few sperm ever

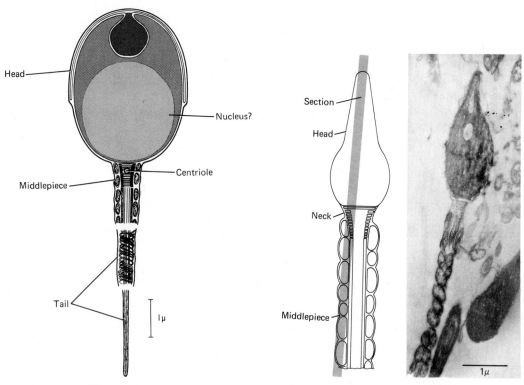

**Figure 5.2**   A human sperm. (*Left*) A "frontal" section (based on electronmicroscopic studies). Sperm diagram from Schultz-Larsen.
(*Right*) Electronmicrograph of a somewhat tangential "longitudinal" section through head and middlepiece. Diagram shows approximate plane of section. Head and middlepiece diagram from Lord Rothschild, *British Medical Journal*, 1 (1958), p. 301.

unite with an egg. Most sperm die. This is due to the use of contraceptives as well as to the fact that intercourse frequently occurs when the woman is not fertile. Also, sperm are ejaculated during activities such as nocturnal emissions and masturbation when there is no possibility of fertilization.

The spermatic fluid in a normal ejaculation has a volume of 2.5–5 cubic centimeters, approximately a teaspoonful. This small amount of fluid contains 150–600 million sperm, which comprise only a very small proportion of the total volume of semen. The ejaculated fluid is whitish and semigelatinous. If a male ejaculates frequently the fluid may have a more watery consistency.

If ejaculation has occurred during intercourse, the sperm will be deposited in the woman's vagina and will begin to make their way to the uterus. One of the first obstacles they face is the force of gravity. If intercourse has taken place while the woman was lying on her back, and if she has remained in that position for sometime afterward, there is a greater chance that some of the sperm will travel toward the cervix and on into the uterus. But if the woman has been in an upright position during intercourse or if she has arisen immediately afterward, the force of gravity will cause the sperm to flow away from the uterus. Even when the woman lies on her back, sperm may be lost through the vaginal opening, especially if this opening has been widened during childbirth. However, as explained in Chapter 3, the engorgement of the blood vessels around the vaginal opening during sexual arousal creates a temporary orgasmic platform that helps keep seminal fluid in the vagina.

The position and structure of the penis may also affect the journey of the sperm. Continued in-and-out thrusting of the penis after ejaculation, for instance, tends to disperse the sperm and hinder them on their journey. The semen also may fail to reach the cervix because of incomplete penetration of the vagina by the penis. This may be unavoidable in the case of extreme obesity or because of a condition known as *hypospadias,* in which the opening of the urethra is located under rather than at the tip of the glans. In such cases artificial insemination may be used to impregnate the woman. In this process the male ejaculates into a container to which a chemical preservative (glycerine) has been added. This fluid is then injected into the vagina or directly into the uterus. Sperm can also be quick-frozen in a glycerine solution and stored for later use.[1]

Even if coitus occurs during a fertile period, there are still many obstacles on the sperm's journey to meet the egg. If the sperm does reach the area of the cervix and is allowed to remain there, it still has another problem to face—acidity. As mentioned in Chapter 4, the pH of cervical secretions varies, and sperm are extremely sensitive to acidity. If the secretions of the cervix and vagina are strongly acidic, the sperm are destroyed quickly. (This is why for contraceptive purposes a vinegar or acetic acid douche is sometimes used to wash out the vagina after intercourse; see Chapter 6). Even in a mildly acidic environment (a pH of 6.0, for instance) the movement of sperm ceases. The best pH for sperm survival and movement is between 8.5 and 9.0 (alkaline). In such an environment the natural tendency of sperm is to swim toward the cervix against the flow of the fluid coming from the cervix.

Sperm swim at a rate of 1 to 2 centimeters (about 1 inch) an hour, but once they are through the cervix and inside the uterus they may be aided in their journey by muscular contractions of the wall of the uterus. Sexual stimulation and orgasm are known to produce contractions of the uterus, a phenomenon of which pregnant women in particular are often acutely aware. Whether or not these contractions aid the sperm in their journey through the uterus is not known for sure; but it is known that sperm may arrive in the fallopian tubes much sooner than would be expected from their own unassisted rate of propulsion—within an hour to an hour and a half after intercourse, in fact.

Once through the uterus and into the fallopian tube, the sperm complete the final 2 inches or so of their journey by swimming against the current generated by small, waving, hairlike structures (cilia) that line the tube. Even if the woman has ovulated, only about half of the sperm that make it through the uterus end up in the fallopian tube that contains the egg (except on rare occasions when a woman ovulates from both ovaries at the same time).

If one considers all the obstacles that sperm face on their journey, it is not surprising that of the several hundred million sperm in the original ejaculation only about 2000 reach the tube that has the egg. And then only one actually unites with the egg.

[1]Sperm may be quick-frozen in a glycerine solution and stored for years. When pregnancy is desired, it is thawed and insemination is performed as described in the text. The use of "sperm banks" as "fertility insurance" for men who have had vasectomies has become somewhat popular in recent years. Present evidence indicates there is at least a 50 percent chance of pregnancy by this method if the sperm is less than five years old. With increasing age, frozen sperm specimens show decreased motility upon thawing. (For further discussion of vasectomy, see Chapter 6).

## The Egg (Ovum)

The egg (ovum) is much larger than the sperm and much larger than the other cells of the body, but it still is quite small compared to the eggs (ova) of other species—scarcely visible to the naked eye (*see* Figure 5.3). It is a spherical cell about 130 or 140 microns in diameter (about 1/175 inch), and weighs about 0.0015 milligram or approximately one-twenty-millionth of an ounce.

The discovery of the human egg has been attributed to Karl Ernst von Baer (1792–1876), generally recognized as the father of modern embryology. Von Baer studied a group of female dogs in various stages of pregnancy and succeeded in finding minute specks of matter in dogs in which the embryos had not had time to develop. He noted that these small objects in the fallopian tubes were smaller than the follicles of the ovaries, which had previously been thought to be the eggs themselves. Von Baer then

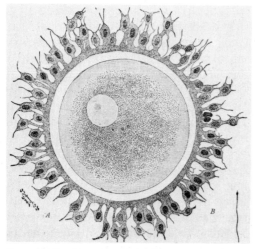

**Figure 5.3** A picture of the human egg, close to maturity. The zona pellucida is the clear circle. The corona radiata is still present in this figure. It is the outer layer of cells. A human sperm is in the lower right-hand corner, showing the relative size of the two cells. From Arey, *Developmental Anatomy*, 7th ed. (Philadelphia: W. B. Saunders Company, 1974.) By permission.

verified that the follicles contained minute bodies identical with those he had previously observed in the fallopian tubes, which had been assumed to represent the actual germ material essential for reproduction. Later studies of humans confirmed these findings.

We noted in Chapter 2 that at birth the ovaries contain about 400,000 primitive egg cells (*primary oocytes*), and it is generally believed that no new egg cells are produced by the human female after birth, in significant contrast to the continuous production of sperm in the male throughout his life. Although only about 400 eggs are essential to provide one a month during a woman's reproductive life, exposure to radiation or drugs may alter all the germ cells at once and do permanent damage.

Possibly related to the "aging" of eggs is the increased incidence of certain defects, particularly Down's syndrome (mongolism), in children born to older women. A woman in her forties is a hundred times more likely to have a Down's child than is a woman of 20. This finding has been clearly correlated with the mother's age, although it is not known what causes chromosome defects as women grow older. For these and other reasons, the age at which it is best for a woman to have a child, biologically speaking, is usually said to be between 20 and 35 years.

### Egg Maturation

An egg matures while still in the ovary, encased in a larger spherical structure known as a *graafian follicle*. The egg divides while still within the follicle, producing a large *secondary oocyte* and a minute *first polar body*, each containing 46 chromosomes. The final step in maturation of the egg occurs after ovulation, within the fallopian tube when division of the genetic material in the cell occurs, resulting in an egg containing 23 chromosomes and a second polar body containing 23 chromosomes. The latter is extruded and disintegrates. Thus the egg contains half of the chromosomes that might be a part of a new being.

The mature egg is a spherical cell containing, in addition to the genetic material (chromosomes), fat droplets, protein substances, and nutrient fluid (see Figure 5.3). It is surrounded by a gelatinous capsule (*zona pellucida*), the final obstacle to the sperm cell seeking to fertilize the egg. (A layer of small cells, the *corona radiata,* that surrounds the egg within the follicle has usually disappeared by this time.)

### Ovulation and Migration of the Egg

The hormonal mechanism that stimulates ovulation has been described in Chapter 4. At the time of ovulation the graafian follicle has reached 10 to 15 millimeters in diameter and protrudes from the surface of the ovary. It is filled with fluid, and its wall has become very thin. The egg has become detached within the follicle and is floating loose in the fluid. At ovulation the thinnest part of the follicle wall finally bursts, and the egg is carried out with the fluid. Although the fluid does build up a certain amount of pressure within the follicle, the process of ovulation is less like the explosion of an air-filled balloon and more like leakage from a punctured sack of water.

Occasionally a follicle fails to mature and rupture; the result is a *follicle (retention) cyst,* one of the most common forms of *ovarian cysts* (a fluid-filled sac in the ovary). Such cysts may vary from microscopic size to 1 to 2 inches in diameter. It is possible for *several* cysts to be present simultaneously in one or both ovaries. If large they may produce pain in the pelvis and cause painful intercourse, and the ovary may be tender to external pressure. Cysts of this kind rarely present serious problems and usually disappear spontaneously within 60 days.

The exact mechanism by which the egg finds its way into the fallopian tube is still a mystery. Although the fringed end of the tube is near the ovary, there is no direct connection that would ensure that the egg did not simply fall into the abdominal cavity (see Figure 2.8). Indeed, it sometimes does, but even more fascinating, it has clearly been established that an egg from an ovary on one side of the body can somehow reach the tube on the other side. There is a well-documented instance of a woman who had her right tube and left ovary removed surgically yet succeeded in having two normal pregnancies afterward.

In any event, once the egg has entered the fallopian tube it begins a leisurely journey to the uterus, taking about three days to move 3 to 5 inches. The egg, in contrast to the sperm, has no means of self-propulsion but is carried along by the current of the small cilia lining the tube. There are also periodic contractions of the muscles in the walls of the fallopian tubes, which are believed by some investigators to aid in transporting the eggs, although this belief has not been definitely confirmed.

If the egg is not fertilized during its journey through the fallopian tube, it disintegrates.

## Fertilization

The moment when sperm and egg unite is the moment of conception, and, though it is commonly believed that conception occurs within the uterus, it actually takes place about halfway down the fallopian tube. If intercourse has taken place within about 72 hours or less before ovulation, viable sperm may be present in the tube, swimming about in apparently random fashion when the egg arrives on the scene. Conception must occur within about 48 hours after ovulation or the egg will be incapable of being fertilized. Assuming that the time of ovulation is known and the woman has intercourse within three days before or two days after this time, she may become pregnant.

Although many sperm surround the egg

## Box 5.2    The Preformationist View

A medieval medical manuscript illustrates the preformationist's beliefs about the fetus's position in the uterus. The preformationists held that the fetus was perfectly preformed from the moment of conception. The only changes that they presumed to occur during pregnancy were growth and shifts in position. The pictures represent the fetus in the uterus as a fully formed human being with adult proportions. (Artists of the period typically drew children as if they were small-scale adults, too.)

at the time of fertilization only one sperm actually penetrates it. This sperm adds its 23 chromosomes to the 23 in the egg, providing the necessary complement of 46 for one fertilized human cell. It is believed that the additional sperm assist conception by secreting a particular enzyme (*hyaluronidase*) that aids in eliminating excess cellular material that may still cling to the outside of the egg.

The mechanism by which a sperm actually penetrates the capsule of the egg is unknown, but apparently it bores through this capsule somehow. In some animals sperm have been observed to enter the eggs and to continue to move well into the interior with the aid of their tails. In other animals only the heads and middlepieces enter the eggs.

By means of another unknown mechanism the wall of the egg becomes impervious to sperm once the first one has successfully penetrated. Whether or not this

sperm represents the "best" of the available specimens is a matter of conjecture; in any event the particular genes that it carries will determine the sex and many other characteristics of the child. If it carries an X chromosome the child will be a girl. If it carries a Y chromosome the child will be a boy.

Attempts to predetermine the sex of offspring are as old as recorded history. Aristotle recommended having intercourse in a north wind if boys were desired, intercourse in a south wind for girls. A more recent formula includes engaging in intercourse two to three days before ovulation and then abstaining if a girl is desired. To conceive a boy the couple must limit intercourse to the time of ovulation. The timing is combined with an acid douche, shallow penetration, and no orgasm for the woman to conceive a girl, or with an alkaline douche, deep penetration, and orgasm for the woman to conceive a boy. While one study found this method to be fairly successful, another study reported conflicting results: Engaging in intercourse several days before ovulation and then abstaining resulted in a higher incidence of boys (not girls); when coitus occurred close to the time of ovulation there was a higher incidence of girls (not boys). These conflicting results reveal problems in the current methods of sex selection.[2] Although there have been numerous attempts to separate X-containing from Y-containing sperm by centrifugation, sedimentation, filtration and other laboratory procedures, none has been notably successful.

A study of 5981 married women in the United States indicates that if sex selection were practiced by this sample, the long-term effect on the ratio of male to female births (currently 105 male to 100 female) would be negligible. However, there was a strong preference for the first child to be male, so that the short-term affect would be a preponderance of male births, followed several years later by a preponderance of female births. When asked their opinions about the actual practice of sex preselection, 47 percent of the women in the sample were opposed, 39 percent were in favor, and 14 percent were neutral.[3]

### Amniocentesis

The most accurate method for finding out the sex of an unborn child is amniocentesis, but this procedure usually is not performed until the fourteenth or sixteenth week after the last menstrual period.[4] Although amniocentesis can be used to find out the sex of a fetus, it is most often used to test the chromosomal or biochemical makeup of the fetus when some abnormality is suspected.

[2]For a further discussion of these methods see Shettles (1972) and Guerrero (1975).
[3]Westoff and Rindfuss (1974).
[4]Researchers are developing several promising techniques for determining the sex of the fetus before 12 weeks of pregnancy.

The procedure of amniocentesis is simple. First, sound waves are used to make a picture (an ultrasonic scan) of the fetus in the womb. Once the position of the fetus is known, it is possible to insert a hollow needle through the abdominal wall into the pregnant uterus without harming the fetus (*see* Figure 5.4). Some of the amniotic fluid that surrounds the fetus (and contains cells of the fetus) is then removed and analyzed. The condition and number of the chromosomes as well as the presence of either an X or Y chromosome can be determined with a high degree of accuracy.

Amniocentesis is a relatively safe procedure for both mother and fetus. In a study sponsored by the National Institutes of Health of more than 1000 amniocenteses there were no major complications. Only three sex determinations and six diagnoses of chromosomal abnormalities were incorrect. About 5 percent of all infants born live have some sort of serious birth defect or will develop mental

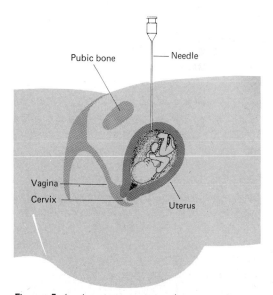

**Figure 5.4**   Amniocentesis involves removing a few milliliters of amniotic fluid to test for cellular abnormalities. Redrawn from *Resident & Staff Physician* (1977). By permission.

retardation. Among the serious abnormalities that can be detected by amniocentesis are Down's syndrome (mongolism), neural tube closure defects (open spine), Tay-Sachs disease, cystic fibrosis, and Rh incompatibility. When abnormalities are diagnosed, women can legally choose to have an abortion—although by the time the chromosome analysis is complete (the twentieth week to twenty-fourth week of pregnancy) abortion is no longer a simple, risk-free operation (*see* Chapter 6).

Because a woman in her forties is 100 times more likely than a woman in her twenties to have eggs with abnormal chromosomes, amniocentesis is currently recommended for all women who become pregnant after the age of 35 and for women who have previously had a child with a chromosome abnormality or with some other problem that can be detected by analysis of fetal cells. It is also recommended for couples with family histories of certain heritable problems. "Genetic counseling" is an important medical obligation, but it can also be distorted into a political and racist weapon.[5]

### Infertility

Even in overpopulated societies failure to conceive is often a problem of great concern to individuals involved. Between 10 and 15 percent of all marriages in the United States are involuntarily childless. Although it is common in many societies to attribute sterility to females, about 40 percent of infertile marriages result from sterility in the male partners. Therefore it is important that, whenever a couple wishes to have children and is unsuccessful for about a year, both the husband and wife should see a physician, for either or both may have disorders that can be treated.

[5]For instance, sickle-cell anemia is an incurable, and often fatal, hereditary disease of the blood that occurs almost exclusively in black people. The racial implications of counseling people with this disease not to have children is obvious, as is the medical dilemma of knowingly perpetuating the illness.

Conditions that interfere with a man's ability to ejaculate into the depths of the vagina are one cause of infertility and have been described earlier in this chapter. More commonly, the problem is a low sperm count, which can result from many factors. An ejaculation that contains fewer than 35 million sperm per cubic centimeter (a total count of fewer than 100–150 million sperm in the seminal fluid) will almost never result in fertilization of the egg. Frequent ejaculation (one or more times a day for several consecutive days) will lower the sperm count; this deficiency, obviously, is easily remedied. Usually, however, the problem is more serious. Infectious diseases that involve the testes may cause sterility. The most common such disease is mumps (if it occurs after puberty). Infections of the prostate gland may also affect fertility. Direct damage to the testes through trauma or radiation may result in decreased sperm production or mechanical obstruction to the movement of the sperm.

*Cryptorchism* (failure of the testes to descend into the scrotum from the abdomen during development) affects about 2 percent of postpubertal boys bilaterally and results in sterility because sperm do not mature at the higher temperature to which they are exposed within the abdomen. This condition can sometimes be remedied either medically or surgically before puberty. If one testicle is present in the scrotum, fertility is usually not impaired.

Certain hormone disorders, particularly hypothyroidism and diabetes, may be associated with low sperm counts. Men with such disorders can often be successfully treated with supplemental hormones.

Another approach to infertility in men with low sperm counts is known as AIH (artificial insemination with husband's semen). This is achieved by collecting and storing a series of ejaculates (as described earlier) and inseminating the wife when sufficient sperm have accumulated. If the husband produces no sperm at all, artificial insemination with donor sperm (AID) is sometimes performed, a procedure which has led to unresolved moral and legal questions. For instance, does AID constitute adultery? In the event of divorce must a man support a child unrelated to him biologically?

The most common cause of infertility in women is failure to ovulate. This may be related to hormonal deficiencies or to other factors, such as vitamin deficiency, anemia, malnutrition, or psychological stress. If a woman does not ovulate, there are several possible solutions. One is the drug *clomiphene,* which stimulates ovulation by causing the release of gonadotrophins from the pituitary. Clomiphene has also been used successfully, on an experimental basis, to treat men with low sperm counts. If a woman does not respond to clomiphene, it is sometimes possible to stimulate ovulation by hormone injections (FSH).

Overstimulation of the reproductive system with hormones or drugs, however, can result in undesirable side effects. Complications include blood clotting and accumulation of fluid in the abdomen. Another problem is multiple births. These occur because it is not always possible to calculate the exact dosage necessary to stimulate one (and only one) follicle. One study of women treated with clomiphene found that 10 percent of them had twins or other multiple births. There also have been cases in which women treated with hormones had pregnancies with as many as 15 fetuses. In such cases the risks of fetal death, miscarriage, and birth complications are extremely high.[6] In at least one case, however, a woman has given birth to sextuplets. The six infants were born live, and the mother did well.

Infections of the vagina, cervix, uterus, tubes, or ovaries may cause infertility by inhibiting or preventing passage of sperm or

eggs to the site of fertilization. Treatment involves identification of the particular microorganism causing the infection and administration of the appropriate antibiotic.

There are also various congenital malformations of the female reproductive tract, as well as tumors (particularly cervical or uterine), that may prevent fertilization. Many of these conditions can be corrected surgically (some conditions are described in more detail in Chapter 7). A woman may inadvertently be preventing conception by using certain commercial douches or vaginal deodorants containing chemicals that destroy or inhibit the motility of sperm.

A physician consulted by a couple having infertility problems will take complete medical histories, and each partner will undergo a physical examination with particular emphasis on the reproductive organs. Laboratory tests to rule out infections, anemia, and hormone deficiencies will be performed. The husband will be instructed to bring in a complete ejaculation for sperm analysis. The wife will be instructed to keep a morning temperature chart during her next several menstrual cycles. The purpose of this chart is to determine if and when she is ovulating, as indicated by a rise in basal body temperature (BBT) approximately in midcycle (the use of such charts to avoid conception in the practice of the "rhythm method" is described in Chapter 6).

If these tests indicate that the male is producing adequate numbers of apparently normal and motile sperm and that the female is ovulating, then further tests to determine why conception is not occurring will be indicated. The most common site of obstruction of the egg in its journey from the ovary is the fallopian tube, and the so-called "Rubin's test" is performed to determine whether the tubes are obstructed. The test involves forcing carbon dioxide under pressure into the uterus through the cervix. If the tubes are open, there will be a drop in pressure as the gas

---

[6] Austin and Short (1972), vol. 5, chapter 4.

passes through them. If there is an obstruction, the pressure will increase.

The most common site of obstruction of the sperm in their journey is the cervix; an analysis of the cervical mucus may also be in order so as to rule out an abnormality at this point. Once a correctable problem has been identified, treatment may begin. If no abnormalities can be found, couples are sometimes advised to adopt children. A previously infertile couple may sometimes succeed in having children of their own after having adopted a child. However, the incidence of pregnancy following adoption is no higher in couples who never adopt, contrary to a widely held view that adoption may "cure" infertility.

### "Test-Tube Babies"

Although there certainly are couples who cannot have children, research into fertility has found solutions to many fertility problems. A method has been developed that enables women who could not become pregnant because of blocked fallopian tubes to conceive. In simplified terms, the egg is removed from the ovaries just prior to ovulation, when the mature follicle is a visible swelling on the ovary. Once removed the egg is mixed with sperm from the husband. After it has been fertilized and has divided into 8 or 16 cells, the conceptus is introduced through the cervix into the uterus. If all goes well, implantation in the endometrium occurs and pregnancy continues normally. There are numerous technical problems, and the failure rate for the procedure is high. In 1978 the first so-called "test-tube baby" was born. This was the first birth to result from a fertilization performed outside the body.

This technique raises many ethical and legal questions. If one accepts the proposition that life begins with conception, then the many experiments that have involved study of fertilized human eggs in the laboratory, in such a way that their ultimate destruction was necessitated, constituted murderous acts; indeed, the scientists involved in this and related research have been accused of just that.

Another possible application of current knowledge in this area would be the implantation of the fertilized egg into the uterus of a surrogate or host mother who would undergo the pregnancy (presumably for a fee) and surrender it to the biological mother at birth. But what if the biological parents have changed their minds, divorced, or simply refuse to accept the child? One law professor has raised these and many other thorny issues that could arise, including such questions as:

What if the host mother refuses to relinquish the child or has intercourse with her husband following implantation . . . and gives birth to twins? Can she be forced to relinquish the child? Both children? If the host mother has an abortion because early months of nausea cause her to change her mind about carrying the fetus to term, would this amount to breach of contract? . . . If she contracted German measles during the first three months of pregnancy so that the child was born deformed, is she responsible for the care of the child or are the couple who contracted with her to produce it responsible?[7]

## Pregnancy

Pregnancy today is an experience which many women look forward to and enjoy. A woman should consult a doctor early in pregnancy (preferably by the second missed period) in order to receive proper care, which can help her have a pregnancy free of complications and deliver a healthy baby. Childbirth has not always been as safe as it is today (see Box 5.4 near the end of the chapter).

The average length of human pregnancy is 266 days, or approximately nine cal-

[7]Schroeder (1974), pp. 542–543. For further discussion of these issues, see also Austin and Short (1972), chapters 4 and 5.

endar months.[8] It is traditional obstetrical practice to divide pregnancy into three-month periods called *trimesters*. We shall discuss each trimester from the perspectives of the expectant mother's experiences and developments within the uterus. The details of each stage in fetal development are available in textbooks on embryology. Our aim is to enable the reader to visualize more clearly the fascinating steps that intervene between conception and birth, without becoming bogged down in details and complex terminology.

## The First Trimester

### First Signs of Pregnancy

Many readers of this book will vividly recall a time when they themselves or acquaintances failed to menstruate as expected. The experience may have been particularly joyful if it signaled the news that mutual desire between a man and woman for a child was to be fulfilled. On the other hand, it may have been an experience of fear and guilt, the unintended consequence of sexual intercourse. Or the news may have been met with mixed feelings, as when a young couple wanted to have children "but not yet." But the knowledge that a woman has missed her period is rarely met with indifference by those concerned.

A woman may miss her period for many reasons other than pregnancy, however, as we have pointed out in Chapter 4. Various

[8]There is great variation from this mean in both directions. Premature births are well known. Extended pregnancies also occur; the length of pregnancy assumes legal importance in establishing the legitimacy of a child when the husband has been away for more than ten months. In the United States the longest pregnancy upheld by the courts as legitimate was 355 days. In England in *Preston-Jones* v. *Preston-Jones,* conducted in the House of Lords in 1949, the husband claimed adultery on the grounds that the interval of sexual abstinence between himself and his wife before birth of the child in question would have meant a pregnancy of 360 days. Although he was initially overruled, he finally won his case in the Court of Appeals.

illnesses and emotional upsets may result in failure to menstruate. In addition, women under 20 and over 40 may skip periods occasionally for no apparent reason. A woman who has recently delivered a child, particularly one who is still nursing, may not menstruate for five or six months or more. She cannot rely on the absence of menstruation as an indication of pregnancy and may indeed become pregnant without having had a menstrual period since her previous pregnancy. She need only have ovulated.

Conversely, a woman may continue to have cyclic bleeding, though in smaller quantities and for shorter durations than usual, during pregnancy. Such bleeding, often called "spotting," occurs in about 20 percent of pregnant women and is not necessarily ominous. It is particularly common among women who have had children before. But "spotting" may also be an early sign of miscarriage. Sometimes such "spotting" has led women who were in the early months of pregnancy to believe falsely that they were not pregnant.

Another early symptom of pregnancy is enlargement and tenderness of the breasts. The physiological mechanisms that lead to enlargement of the breasts have been described in Chapter 4. When hormonal stimulation of the mammary glands begins after conception, a woman will initially become aware of sensations of fullness and sometimes of tingling in her breasts. The nipples in particular become quite sensitive to tactile stimulation early in pregnancy.

Many women experience so-called "morning sickness" during the first six to eight weeks of pregnancy. Usually it consists of queasy sensations upon awakening, accompanied by an aversion to food or even to the odors of certain foods. This nausea may be accompanied by vomiting and great reluctance to be near food. Some women experience "morning sickness" in the evening, but about 25 percent of pregnant women never experience any vomiting at all. On the other hand, about one in 200 pregnant women in the United States experiences vomiting so severe that she must be hospitalized. This condition is known as *hyperemesis gravidarum* and may have serious consequences (including malnutrition) if it is not properly treated. Occasionally therapeutic abortion may be required in order to save the life of the mother, but this necessity is quite rare.

A phenomenon known as "sympathetic pregnancy" is sometimes observed in the husband of a pregnant woman. He too becomes nauseated and vomits along with his wife. The etiology of this condition is implied in its name.

More frequent urination is another symptom of early pregnancy and is related to increased pressure on the bladder from the swelling uterus. This symptom tends to disappear as the uterus enlarges and rises up into the abdomen, but frequent urination again becomes a common complaint toward the end of pregnancy when the fetal head descends into the pelvis and exerts pressure on the bladder.

Fatigue and a need for more sleep are often most striking early in pregnancy and may be quite puzzling to a woman who is usually very energetic. The sensation of drowsiness early in pregnancy may be so overwhelming as to make sleep irresistible, even in the midst of conversation.

### Early Diagnosis of Pregnancy

A sexually active woman who has missed her menstrual period, is having "morning sickness," and has noticed breast enlargement and tenderness, increased frequency of urination, and extreme fatigue has reason to suppose that she is pregnant. If she is particularly anxious to verify this suspicion, she may see a physician for a "pregnancy test." Actually there are many ways in which a physician can determine whether or not a woman

is pregnant, none of which is 100 percent accurate early in pregnancy.

The sixth week of pregnancy (that is, about four weeks after missing a period) is the earliest point at which a physician can by physical examination, determine with any reliability that a woman is pregnant. By this time certain changes in the cervix and uterus are apparent upon pelvic examination. Of particular usefulness to the physician is *Hegar's sign,* which refers to the soft consistency of a compressible area between the cervix and body of the uterus and becomes apparent in the sixth week of pregnancy. To an experienced clinician it is a fairly reliable indication. The examination is performed by placing one hand on the abdomen and two fingers of the other hand in the vagina.

Various laboratory tests can also be performed to detect pregnancy in its early stages. If performed correctly these tests are 95 to 98 percent accurate. The most often used such tests are based on the presence of HCG (a hormone produced only by pregnant women). These tests are relatively simple and inexpensive. The woman's blood or urine sample is put into a test tube or in a small dish with specific chemicals. If HCG is present, it can be detected within minutes, even in very small amounts. The test for HCG using the woman's urine is the most common test for determining pregnancy. It can also detect HCG as early as two weeks after conception. The blood test for HCG is used less frequently in part because it is a more expensive test. The advantage of the blood test is that it can detect HCG within six to eight days after fertilization.

A relatively inexpensive version of the HCG test is also available for home use. If used correctly, it can detect HCG in the urine of a pregnant woman as early as nine days after a missed period. Recent studies found that only 3 percent of home tests indicated pregnancy when the women were not pregnant. But, in 20 percent of tests giving negative results, the women were actually pregnant even though the tests indicated that they were not. When the women with negative results performed a second test eight days later, the accuracy of the test increased to 91 percent.[9] Thus, home tests giving positive or negative results should be confirmed by a doctor.

Once pregnancy has been reasonably established, the first question that the woman asks is usually, "When is the baby due?" The expected delivery date (EDC, or expected date of confinement) can be calculated by the following formula: Add one week to the first day of the last menstrual period, subtract three months, then add one year. For instance, if the last menstrual period began on January 8, 1980, adding one week (to January 15), subtracting three months (to October 15), and adding one year gives an expected delivery date of October 15, 1980. In fact, only about 4 percent of births occur on the dates predicted by this formula. But 60 percent occur within five days of dates predicted in this manner.

### False Pregnancy (Pseudocyesis)

Occasionally a woman may become convinced that she is pregnant despite evidence to the contrary. About 0.1 percent of all women who consult obstetricians fall into this category. Usually they are young women who intensely desire children, but some are women near menopause. A certain percentage of these women are unmarried; in fact, the phenomenon was not unknown among nuns in medieval convents, who took their marriage vows with Christ literally and believed that they had been impregnated by Christ.

Women suffering false pregnancy often experience the symptoms of pregnancy, including morning sickness, breast tenderness, a sense of fullness in the pelvis, and the sensation of fetal movements in the abdomen.

[9]McQuarrie and Flanagan (1978).

They often cease to menstruate, and physicians may observe contractions of the abdominal muscles that resemble fetal movements. Even though pregnancy tests are negative, a woman suffering from a severe mental illness like schizophrenia may persist in her delusion for years.

### Ectopic Pregnancy

An ectopic pregnancy occurs when a fertilized egg implants outside of the uterus. About 1 in 250 pregnancies is ectopic. Often such pregnancies spontaneously abort. The most common site of ectopic pregnancy is the fallopian tubes. If an embryo that implants in the tube continues to develop, the tube may rupture, causing hemorrhaging and in some cases death to the mother. Signs of an ectopic tubal pregnancy are lack of menstruation, abdominal pain, and nonmenstrual bleeding.

Rare sites for ectopic pregnancies include such locations as the ovary or intestines. In very rare cases the pregnancies result in the birth of a child (via surgical methods). In 1979 a New Zealand woman who previously had had her uterus removed gave birth during surgery to a baby girl after eight months of pregnancy. The fertilized egg had become attached to her bowel, where it continued to develop.

### Sexual Activity during the First Trimester

There is no reason why early pregnancy or the suspicion of pregnancy should inhibit a healthy woman's sexual activities. Although occasional morning sickness may cause lack of interest during the early part of the day, and fatigue may be a deterrent in the evening, most couples are able to find mutually satisfactory occasions for continued sexual relations. Decreased sexual interest was reported by about one-fourth of women in a recent study, but the frequency of sexual activity did not change significantly when compared to frequency prior to conception (*see*

**Table 5.1** Percent of Women Having Various Frequencies of Coitus at Different Stages of Pregnancy

| Number of acts of coitus per week | Baseline (1 year before conception) | 1st trimester | 2nd trimester | 7th month | 8th month | 9th month |
|---|---|---|---|---|---|---|
| None | 0% | 2% | 2% | 11% | 23% | 59% |
| 1 | 7% | 11% | 16% | 23% | 29% | 19% |
| 2–5 | 81% | 78% | 77% | 63% | 46% | 23% |
| 6 or more | 12% | 9% | 5% | 2% | 2% | 1% |

Table 5.1).[10] Factors influencing sexual interest in the first trimester include nausea, fatigue, and probably general anxiety in some women who are pregnant for the first time or who had complications with a previous pregnancy. Also, physiological changes during this period can make sexual stimulation somewhat painful. In particular, the vasocongestion of the breasts that occurs during sexual excitement may be painful to a woman who is already experiencing tenderness there.

There is no evidence to support the notion that intercourse during the early months of pregnancy can cause abortion of the fetus. Nevertheless, concern about injuring the fetus during intercourse is often reported by women during the first trimester, particularly if they are pregnant for the first time or have had previous miscarriages as has been said. It is interesting that some preliterate societies believe continued intercourse after conception essential to the continuation of pregnancy; they believe that the sperm serve to nourish the developing fetus.

### Intrauterine Events of the First Trimester

After the human egg has been fertilized, it begins to divide into multiple cells as it moves down the fallopian tube (see Figure 5.5). The original egg becomes two cells, these two become four, the four become eight, and so on. There is no significant change in volume during these first few days,

[10]Wagner and Solbert (1974).

but the initial egg has become a round mass subdivided into many small cells. This round mass of cells is called a *morula* (from the Latin *morum,* "mulberry"), and the cells of the morula arrange themselves around the outside of the sphere during the third to fifth days after ovulation, leaving a fluid-filled cavity in the center. This structure, called a *blastocyst,* floats about in the uterine cavity, and sometime between the fifth and seventh days after ovulation attaches itself to the uterine lining and begins to burrow in, aided by enzymes that digest the outer surface of the lining, permitting the developing egg to reach the blood vessels and nutrients below (see Figure 5.6). By the tenth to twelfth day after ovulation the blastocyst is firmly implanted in the uterine wall, yet the woman still cannot know that she is pregnant, for her menstrual period is not due for several more days.

The blastocyst, which has embedded itself in the lining of the uterus, will develop from a tiny ball of cells into an easily recognizable human fetus during the first trimester. In the early stages of development a disk-shaped layer of cells forms across the center of the blastocyst and from this *embryonic disk* the fetus grows. The remaining cells develop into the *placenta,* the *membranes* that will contain the fetus and the *amniotic fluid,* and the *yolk sac* (which in humans is insignificant in function).

### *Development of the Placenta*

The placenta is the organ that exchanges nutrients and waste products with

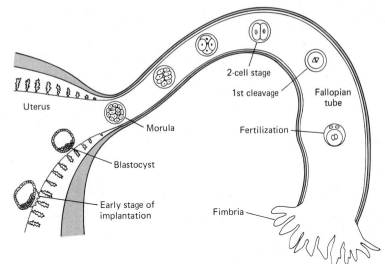

**Figure 5.5** Fertilization to implantation of the human embryo. From Tuchmann-Duplessis and Haegel, *Illustrated Human Embryology.* vol. I (New York: Springer-Verlag; London: Chapman and Hall; Paris: Masson et Cie, 1971). Reprinted by permission.

the mother. It is constructed from both fetal and maternal tissue and during the first trimester develops into a bluish-red, round, flat organ about 7 inches (18 centimeters) in diameter and 1 inch (2.5 centimeters) thick. It weighs about 1 pound (0.45 kilogram) and constitutes, along with the fetal membranes, the "afterbirth." The blood vessels of the placenta are connected to the circulatory system of the mother through the blood vessels in the wall of the uterus. The circulatory system of the fetus is connected to that of the placenta by the blood vessels of the *umbilical cord,* which is attached to the placenta. Ox-

ygen and essential nutrients reach the fetus through the umbilical vein, and waste products from the fetus reach the maternal system through the two umbilical arteries.

The placenta also functions as an endocrine gland, producing hormones essential to the maintenance of pregnancy. The hormone human chorionic gonadotrophin (HCG) stimulates continued production of progesterone by the corpus luteum during the first trimester.[11] Gradually, however, the placenta

[11]Claims that administering additional doses of HCG by injection during pregnancy will raise a baby's IQ have no basis in fact.

**Figure 5.6** Low- and high-power photographs of the surface of an early human implantation obtained on the twenty-second day of cycle, less than eight days after conception; the site was slightly elevated and measured 0.36 × 0.31 mm. (*Left*) the actual size of the embryo indicated by the white square at the lower right; the mouths of uterine glands appear as dark spots surrounded by halos. From Eastman and Hellman, *Williams Obstetrics,* 13th ed. (New York: Appleton-Century-Crofts. 1966), p. 123. Reprinted by permission.

**Box 5.3**   Cojoined Twins

Congenital malformations have always fascinated the human mind. In early days, malformed babies were called monsters. In a book written in 1560, Ambroise Paré wrote, "We called Monsters, what things soever are brought forth contrary to the common decree and order of nature."[1] Cojoined twins were well known at the time, even though this malformation is quite rare (approximately 1 in 50,000 births). According to Paré, the most likely cause of cojoined twins was thought to be "An abundance of Seed and overflowing of matter."

In a book titled *Cosmology*, published in 1552, Sebastian Munster described having seen two girls in Mainz who were joined at the forehead. He presented the following explanation. When the mother was pregnant with the two daughters, she had been gossiping with

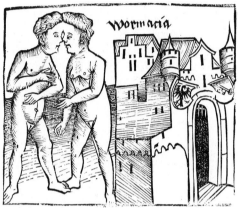

Cojoined twins similar to those described by Munster. An unsigned woodcut from 1510 in the collection of the Yale Medical Historical Library.

[1]Guttmacher and Nichols (1967).

itself begins to produce large amounts of progesterone and estrogens. It also produces other hormones, like cortisol and androgens, in small amounts. There is evidence that placental production of progesterone and estrogens falls off just before delivery and that this drop plays a role in initiating labor.

*Development of the Fetus*

The first trimester is a period during which the rather simple structure of the embryo is transformed into the very complex organism called the *fetus*, though with relatively little change in size. The embryo becomes a fetus after the eighth week of pregnancy. (The ovum becomes an embryo one week after fertilization.)

The embryonic disk described earlier becomes somewhat elongated and ovoid by the end of the second week after fertilization. The embryo is about 1/16 inch (1.5 milli-meters) long at this point. The actual sizes of the embryo at various stages in the first seven weeks are depicted in Figure 5.7.

During the third week growth is particularly noticeable at the two ends of the embryo. Differentiation of the *cephalic* (head) end is particularly notable in comparison to that of the rest of the body. By the end of the third week or the early part of the fourth week the beginnings of eyes and ears are visible. In addition, the brain and other portions of the central nervous system are beginning to form. By the end of the fourth week two bulges are apparent on the concave (front) side of the trunk. The upper one represents the developing heart and is called the *cardiac prominence;* the lower one, the *hepatic prominence,* is caused by the protrusion of the developing liver (*see* Figure 5.8). At this point the embryo is only about four millimeters long and weighs about 1/7 of an ounce

## Box 5.3   continued

another woman and unexpectedly their two heads struck together. The pregnant woman became ill with fright and the fruit within her womb suffered the consequences.

Today it is recognized that cojoined twins represent an aberration in the process that leads to the development of identical twins.[2] Cojoined twins result from a partial but incomplete splitting of a single embryo into two separate twins. The area at which they are joined can lie at any point along their bodies: the abdomen, chest, back, or top of the head.

Some of the most well known of cojoined twins in modern times are the Siamese twins, Chang and Eng, who were born in 1811 in Siam (now Thailand) of Chinese parents. They were joined by a small band of skin and were not greatly impaired in movement. They were displayed in P. T. Barnum's circus for many years.

Surgical procedures can be successful in separating cojoined twins if the twins do not share vital organs.

Chang and Eng Bunker. From an early description by J. C. Warren in 1929. (An account of the Siamese twin brothers united together from birth. *American Journal of Medical Science*, 1829).

[2]Zimmermann (1967).

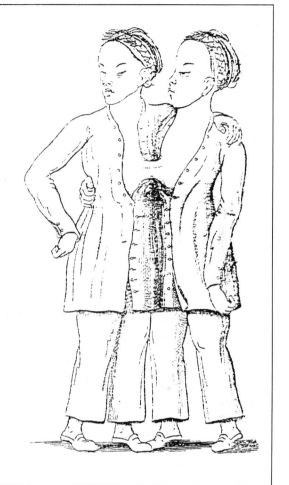

(0.4 gram). A prominent "tail" is present in the young embryo, but it is almost gone by the eighth week.[12]

Between the fourth and eighth weeks the facial features—eyes, ears, nose, and mouth—become clearly recognizable (*see*

[12]In rare instances the tail fails to regress, and there are documented cases of human infants born with tails. The tails are usually removed surgically at an early age but in one case reported in the medical literature a 12-year-old boy had a tail 9 inches long.

Figure 5.9). Fingers and toes begin to appear between the sixth and eighth weeks (*see* Figure 5.10). Bones are beginning to ossify, and the intestines are forming. By the seventh week the gonads are present, but cannot yet be clearly distinguished as male or female. Similarly, the external genitals cannot be identified as male or female until about the third month (*see* Chapter 2).

Between the eighth and twelfth weeks the fetus increases in length from about 1.5

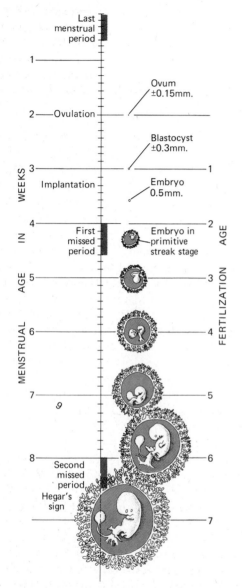

**Figure 5.7** Actual sizes of embryos and their membranes in relation to a time scale based on the mother's menstrual history. From Patten, *Human Embryology*, 3rd ed., p. 145. Copyright © 1968 by McGraw-Hill, Inc. Used with permission of McGraw-Hill Book Company.

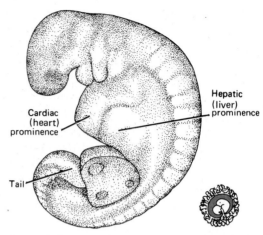

**Figure 5.8** A human embryo about four weeks after fertilization (crown-rump length 3.9 mm). Retouched photograph (×20) of embryo 5923 in the Carnegie Collection. Sketch, lower right, shows actual size of embryo. As modified by Patten for *Human Embryology*, 3rd ed., p. 70, McGraw-Hill Book Company. With permission.

on the pinkish color and the internal genitals become recognizable as male or female. Although still very small, the fetus at 12 weeks is unmistakably human, and from this point on development consists primarily of enlargement and further differentiation of structures that are already present.

### Complications during the First Trimester

Miscarriage, or *spontaneous abortion*, occurs most often during the first trimester of pregnancy. Between 10 and 15 percent of all pregnancies end in spontaneous abortion; that is, the pregnancies terminate in miscarriages before the fetuses have any chance of survival on their own. About 75 percent of spontaneous abortions occur before the sixteenth week of pregnancy, and the great majority occur before the eighth week. The first sign that a woman may miscarry is vaginal bleeding, or "spotting," which was also mentioned earlier. If the symptoms of pregnancy

inches to about 4 inches, and in weight from about 1/15 ounce (2 grams) to approximately 2/3 ounce (about 19 grams). The skin takes

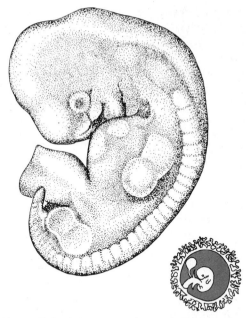

**Figure 5.9**  A human embryo a little more than six weeks after fertilization (crown-rump length 14 mm). Retouched photograph (× 8) of embryo 1267A in the Carnegie Collection. From *Human Embryology*. 3rd ed., p. 74, Copyright © 1968 by McGraw-Hill, Inc. Used with permission of McGraw-Hill Book Company.

disappear and the woman develops cramps in the pelvic region, the fetus is usually expelled. About 50 percent of such miscarried fetuses are clearly defective in some way. About 15 percent of miscarriages are caused by illness, malnutrition, trauma, or other factors affecting the mother. In the remaining 85 percent the reasons are not apparent. A woman who has had one spontaneous abortion usually can conceive again and have a normal pregnancy.

*Effects of Drugs and Diseases*

It is very important to realize that substances other than nutrients may reach the developing fetus through the placental circulatory system. Various drugs taken by the mother may have harmful effects on the developing fetus, particularly during the first trimester. Therefore, a woman should avoid taking any drugs (including aspirin and vitamins) during pregnancy without first consulting her doctor.

In recent years the effects of the sedative thalidomide have been widely publicized. This drug causes abnormal development of

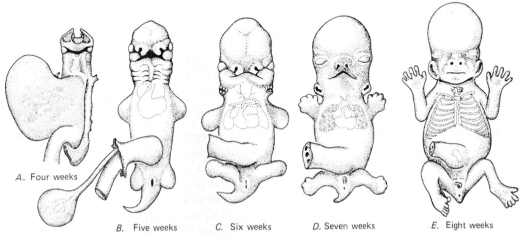

A. Four weeks    B. Five weeks    C. Six weeks    D. Seven weeks    E. Eight weeks

**Figure 5.10**  Frontal views of a series of human embryos, drawn as they would appear if the body curvatures had been straightened. From Patten, *Human Embryology*, 3rd ed., p. 146. Copyright © 1968 by McGraw-Hill, Inc. Used with permission of McGraw-Hill Book Company.

the arms and legs of the baby (*phocomelia*). In infants suffering from this condition the hands and feet are attached to the bodies by short stumps rather than by normal arms and legs.

In the past 20 years a number of women were treated with various combinations of sex hormones during the first trimester to prevent threatened miscarriage. Follow-up studies have shown that girls whose mothers received androgenic hormones during the first trimester may exhibit alterations in behavior as well as in anatomy. The clitoris may be enlarged at birth, and the behavior of these girls, when compared with matched controls who were not exposed to the hormones, was distinctly "tomboyish." They showed a preference for athletic activities as opposed to playing with dolls, for example.[13] Boys whose mothers received exogenous female hormones (estrogen and progesterone) have been found to rank significantly lower on scales of masculinity, assertiveness, and aggressiveness.[14]

A potent estrogen known as DES (diethylstilbestrol) has been shown to cause cancer of the vagina in some girls whose mothers took the drug while pregnant. DES is still in use as the "morning-after pill" (*see* Chapter 6).

Regular use of addictive drugs, like heroin and morphine, by the mother produces addiction in the fetus. When the infant is born, it must be given further doses of the drug to prevent withdrawal effects that may be fatal.

Alcohol is another drug that can cause birth defects. Researchers have found that facial, limb, and heart defects are more common among children of women who drink heavily during pregnancy than among those of women who do not drink. Three ounces (88.7 milliliters) of alcohol (for example, whis-

key) a day or one big binge may be enough to cause damage.

Smoking during pregnancy is harmful. Fetuses carried by women who smoke while pregnant have an increased risk of prematurity, low birth weight, and death before birth than fetuses carried by women who do not smoke. Also, newborn infants born to women who smoke during pregnancy have a greater risk of death than babies born to women who do not smoke.[15]

Certain viruses are also known to have damaging effects on the fetus during the first trimester. The most common is the virus of rubella, or German measles. If a woman has rubella during the first month of pregnancy, there is a 50 percent chance that the infant will be born with cataracts, congenital heart disease, deafness, or mental deficiency. In the third month of pregnancy the risk of such abnormalities decreases to 10 percent. Since 1969 a rubella vaccine has been available in the United States. About 85 percent of adult women have acquired immunity to rubella since they had it as children. If she is in doubt about her immunity, a woman should first be tested and, if found to be susceptible, vaccinated. However, a woman should *not* receive rubella vaccine if she is already pregnant and should avoid conception for six months after the vaccination because the vaccine contains the live rubella virus, which could cause the disease it is designed to prevent. Rubella is different from "regular measles," and being vaccinated against the latter does not protect a person from rubella.

A potentially serious and even fatal anemia of the fetus can result from the transfer across the placenta of antibodies to the red blood cells of the fetus. These antibodies occur when the mother is Rh negative and the fetus is Rh positive. The disorder is called Rh incompatibility. For the fetus to be at risk in this disease, the mother must be Rh neg-

[13]Erhardt and Money (1967).
[14]Yalom et al. (1973).

[15]Meyer and Tonascia (1977).

ative and the father Rh positive (as is the case in about 10 percent of U.S. marriages). Normally, Rh incompatability is not a problem in the first pregnancy. The mother develops antibodies if Rh positive fetal blood mixes with her Rh negative blood. This most commonly occurs during childbirth, miscarriage, or abortion. In the mother's next pregnancy with an Rh positive fetus her antibodies can travel across the placenta and attack the red blood cells of the fetus. This happens in about one out of every 200 pregnancies. In most cases the serious results of this disease (death or brain damage) can be avoided by early identification (parental blood-typing and, in women with a previous history of the disease, amniocentesis) and prompt treatment (blood transfusion of the fetus). The incidence of Rh incompatibility has been reduced due to the development of a vaccine, Rhogam. After an Rh negative woman miscarries, aborts, or delivers, she is given an injection of Rhogam within 72 hours, which keeps her from developing the harmful antibodies. Matings of two Rh positive or two Rh negative individuals do not lead to Rh incompatibility, nor does the mating of an Rh positive woman and an Rh negative man.

### Positive Diagnosis of Pregnancy

From a medicolegal point of view, pregnancy can be positively established by one of three means: (a) verification of the fetal heartbeat, (b) photographic demonstration of the fetal skeleton, (c) observation of fetal movements. Until very recent developments in technology none of these signs of pregnancy could be verified until well into the second trimester. Using a conventional stethoscope a physician can hear the fetal heartbeat by the fifth month. The fetal heart rate is 120–140 beats per minute, and this can be differentiated from the mother's heartbeat, which is usually 70–80. A fetal pulse detector now available commercially can detect the fetal heartbeat as early as 9 weeks, and reliably

after 12 weeks. This device is based on ultrasonic technology. A sound wave of very high pitch is directed at the uterus and is reflected back to a receiver. Movements of the fetal heart cause changes in the pitch of the reflected sound, which are converted to an audible tone and amplified.[16]

Variations in the echo from an ultrasonic pulse resulting from reflection off the fetal skeleton can now be converted to a photographic image that is far more distinct than a conventional X ray. Present evidence indicates that the ultrasonic technique is safer than X rays since it does not involve the radiation hazard to the fetus inherent with X rays. Obviously, neither technique is used routinely, but each is of value in verifying suspected complications such as fetal head size larger than the pelvic opening, retarded fetal head size secondary to a malfunctioning placenta, gross malformation, and multiple fetuses.

Fetal movements can usually be felt by the end of the fifth month. Kicking movements of the fetus become outwardly visible later in the second trimester. Fetal movements not only confirm pregnancy but indicate that the fetus is alive.

## The Second Trimester

### Experiences of the Mother

The pregnant woman becomes much more aware of the fetus during the second trimester because of fetal movements ("quickening") just described. In addition, her waistline begins to expand (particularly in women who have borne children previously), and her abdomen begins to protrude, necessitating a change from regular to maternity clothes. Her pregnancy now becomes publicly recognizable.

The second trimester is generally the most peaceful and pleasant period of preg-

[16]Goodlin (1971).

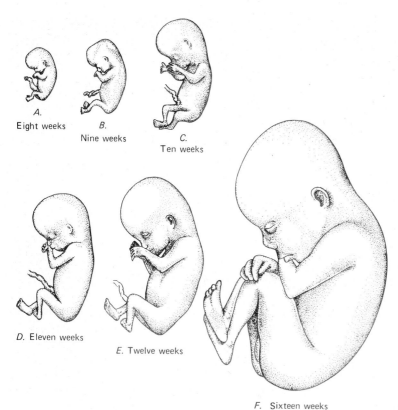

A.
Eight weeks

B.
Nine weeks

C.
Ten weeks

D. Eleven weeks

E. Twelve weeks

F. Sixteen weeks

**Figure 5.11** Human fetuses between 8 and 16 weeks, about one-half actual size. *A-D* from photographs of embryos in University of Michigan Collection. *E* and *F* are redrawn, with slight modification, from DeLee. "Obstetrics." From Patten, *Human Embryology*, 3rd ed., p. 150. Copyright © by McGraw-Hill, Inc. Used with permission of McGraw-Hill Book Company.

nancy. Nausea and drowsiness present during the first trimester tend to disappear during the second. Concerns about miscarriage are generally past, and it is too early to start worrying about labor and delivery. Barring complications or illness, the pregnant woman can be quite active during the second trimester. She can continue to work if she wishes,[17] do housework, travel, and participate in recreation and sports. Rather than eliminating specific activities, most physicians nowadays urge "moderation in all things."

[17]Some employers still discharge pregnant employees or require them to take maternity leave. These practices are being challenged in the courts and will probably be eliminated as discriminatory in most instances. Yet in 1974 the U.S. Supreme Court ruled that pregnancy is not a disability, and therefore a woman who must stop working because of pregnancy is not eligible for disability benefits.

### Sexual Activity during the Second Trimester

Frequency of sexual intercourse does not decrease significantly during the second trimester (*see* Table 5.1). Some women who experienced extreme nausea and fatigue during the first trimester may have renewed sexual interest as these symptoms disappear. There is usually an increase in vaginal lubrication from the second trimester on—and less breast tenderness. However, the expanded abdomen of the later stages of pregnancy produces a pronounced shift in preferred coital positions during the second and third trimester. Use of the male superior position declines significantly and the female superior, side-by-side, and rear-entry positions are used more frequently. Frequency of orgasm declines somewhat during the second trimes-

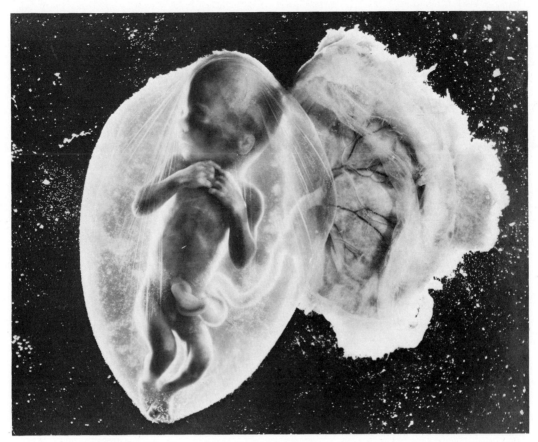

**Figure 5.12**   Fetus, about four months, eight inches long. From Nilsson et al., *A Child Is Born* (Boston: Delacourte Press Seymour Lawrence, Inc. 1965), pp. 116–117. Reprinted by permission.

ter, and, unlike coital frequency, is closely associated with level of sexual interest on the part of the woman.[18]

### Development of the Fetus

Beyond the twelfth week the fetus is clearly recognizable as a human infant (*see* Figures 5.11 and 5.12), although it is much smaller than it will be at birth. It has a proportionately large head with eyes, ears, nose, and mouth. The arms and legs, which began as "limb buds" projecting from the trunk, now have hands and feet, fingers and toes.

The digits of the hands and feet begin with the formation of four radial grooves at the ends of the limb buds, lending the initial appearance of webbed hands or feet (*see* Figure 5.13). In some children remnants of this webbing remain after birth, causing embarrassment to the parents, perhaps, but it usually is of no other significance.

Hair appears on the scalp and above the eyes in the fifth or sixth month. The first hairs, called *lanugo*, are fine, soft, and downy. The skin at the fifth and sixth month is quite red because the many small blood vessels below its surface show through. Beginning in the seventh month, however, layers of fat

[18]Wagner and Solberg, op. cit.

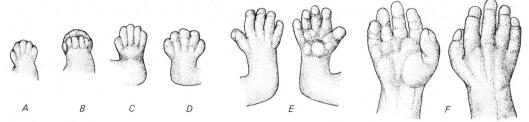

**Figure 5.13**  Stages in early development of the hand. (*A*) Anterior limb bud of an embryo 12 mm long; (*B*) anterior limb bud of an embryo 15 mm long; (*C*) anterior limb bud of an embryo 17 mm long; (*D*) hand and forearm of an embryo 20 mm long; (*E*) two views of the hand and forearm of an embryo 25 mm long; (*F*) two views of the hand of a fetus 52 mm long. After Retzius, from Scammon, in Morris, *Human Anatomy.* From Patten, *Human Embryology,* 3rd ed., p. 148. Copyright © 1968 by McGraw-Hill, Inc. Used with the permission of McGraw-Hill Book Company.

build up beneath the skin, and the baby develops the characteristic chubby pinkness.

Beside further maturation of the internal organ systems during the second trimester, there is a substantial increase in the size of the fetus. At the end of the third month the fetus weighs about 1 ounce and is only about 3 inches long. By the end of the sixth month it weighs about 2 pounds and is about 14 inches long. By this time the eyes can open, and the fetus moves its arms and legs spontaneously and vigorously. The fetus alternates between periods of wakefulness and sleep. The uterus is a very sheltered environment, but such stimuli as a loud noise near the uterus, a flash of high-intensity light, or a rapid change in the position of the mother can disturb the tranquility of the womb and provoke a vigorous movement by the fetus. Changes in outside temperature are not perceived by the fetus, since the intrauterine temperature is maintained by internal mechanisms at a very constant level just slightly (0.5°C) above that of the mother.[19]

If delivered at the end of six months, the fetus usually survives for a few hours to a few days. With heroic efforts on the part of the medical staff 5 to 10 percent of babies weighing about 2 pounds (800 to 1000 grams) survive. The smallest infant known to have survived weighed less than a pound at birth (it weighed 400 grams; 1 pound = 455 grams). The fetus was estimated to be 20 weeks old at the time of delivery. Since fetuses can now be legally aborted up to 24 weeks, it is possible some could survive with extraordinary medical assistance. This is only one of many facets of the ongoing abortion controversy (*see* Chapters 6 and 15).

## The Third Trimester

### Experiences of the Mother

During the last three months of pregnancy the expectant mother becomes more acutely aware of the child that she carries within her swelling abdomen (*see* Figure 5.14). The fetus becomes quite active, and its seemingly perpetual kicking, tossing, and turning may keep the mother awake at night. The woman's weight, if not controlled up to this point, may become a problem, as the woman realizes that she has already gained her "quota" and still has three months to go. Often working against her is an increased appetite that is partly caused by the hormonal changes of pregnancy. According to the Committee on Maternal Malnutrition of the National Research Council, the desirable

[19]Austin and Short (1972), vol. 2, chapter 3.

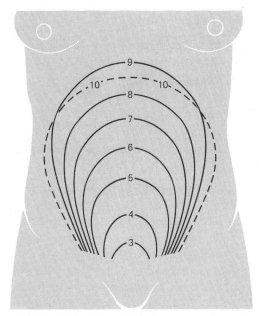

**Figure 5.14** Relative height of the top of the uterus at the various months of pregnancy. From Eastman and Hellmann, *Williams Obstetrics*, 13th ed. (New York: Appleton-Century-Crofts, 1966), p. 263. Reprinted by permission.

weight gain during pregnancy is 24 pounds (10.9 kilograms). The average infant at nine months weighs only about 7.5 pounds (3.4 kg). The rest of the weight gain is accounted for as follows: the placenta, about 1 pound (0.4 kg); the amniotic fluid in the uterine cavity, 2 pounds (0.9 kg); the enlargement of the uterus, 2 pounds (0.9 kg); enlargement of the breasts, about 1.5 pounds (0.7 kg); and the retained fluid and fat accumulated by the mother, the remaining 10 pounds (4.5 kg) or more. There are several reasons why attention is paid to the mother's weight gain during pregnancy. Most important is the higher incidence of medical complications during pregnancy (for example, strain on the heart and high blood pressure) in women who gain excessively. In addition, movement becomes more awkward for the woman who has gained perhaps 40 pounds (18 kg), and she tires more easily.

By the ninth month of pregnancy a woman is usually anxious to "have it over with" and "see what it is." There is speculation about the sex of the child; prospective names are considered; and the nursery is made ready. As the delivery date approaches, a woman may also become anxious about whether or not the baby is going to be "all right"—healthy and without congenital defects.

### Prematurity

A major complication during the third trimester is premature labor and delivery of the fetus. Since the date of conception is not always accurately estimated, and because the age and weight of the fetus are highly correlated, it is traditional to define prematurity by weight rather than estimated age. An infant who weighs less than 5 pounds, 8 ounces (2.5 kg) at birth is considered premature. The mortality rate among premature infants is directly related to size: The smaller the infant, the poorer are its chances for survival. Although an infant born in the seventh month or later can usually survive for a few hours without great difficulty, a smaller one, between 2 and 4 pounds (0.9 and 1.8 kg), often develops difficulty in breathing leading to severe respiratory distress and death within 48 hours.

It is estimated that about 7 percent of births in the United States are premature. Prematurity may be associated with various maternal illnesses—like high blood pressure, heart disease, and syphilis—or with factors such as heavy cigarette smoking or multiple pregnancy. At least 50 percent of the time, however, the cause of prematurity is not clearly known.

### Complications during the Third Trimester

One of the most serious complications of pregnancy, if untreated, is a condition called *toxemia* (from the Latin word for "poison''). In fact, much modern prenatal care has been developed as a result of research done on the cause and treatment of toxemia. The cause of toxemia is still unknown, but it seems that a toxin, or poison, produced by the body causes the symptoms—high blood pressure, protein in the urine, and the retention of fluids by the body. This disease occurs only in pregnant women, usually in the last trimester, and if it is not treated successfully the result can be death for both the mother and the child. Uncontrolled toxemia is a major cause (along with hemorrhage and infection) of maternal mortality. Between 6 and 7 percent of all pregnant women in the United States develop toxemia and a small percentage of these die.

### Sexual Activity during the Third Trimester

Frequency of sexual intercourse (and orgasm) declines significantly with each successive month of the third trimester (*see* Table 5.1). The most common reasons given for this change are, in order of frequency: physical discomfort, fear of injury to the baby, loss of sexual interest, awkwardness in having sexual relations, recommendation of a physician, and feelings of loss of attractiveness (as perceived by the woman).[20]

The primary medical concern about intercourse and/or orgasm late in pregnancy is the risk of inducing premature labor. It is well known that orgasm in the third trimester is accompanied by strong contractions of the uterus, and in the ninth month the uterus may go into spasm for as long as a minute following orgasm. One study reported a 15 percent risk of premature labor or rupture of fetal membranes in women who were orgasmic after the thirty-second week.[21] (Rupture of the fetal membranes can lead to infection as

well as prematurity.) It would appear that the women most at risk are those with a prior history of vaginal bleeding during pregnancy, ruptured membranes, or a prematurely ''ripe'' (soft) cervix. Women who do not have a history of these complications and who are otherwise healthy can probably participate in sexual activity throughout the third trimester at no risk to themselves or the fetus. Some couples prefer to practice intracrural (between the thighs) intercourse or mutual masturbation during the last month of pregnancy.

### Final Development of the Fetus

The third trimester is a period of growth and maturation for the fetus, which by now has developed all the essential organ systems. By the end of the seventh month the fetus is about 16 inches (40.6 cm) long and weighs about 3 pounds, 12 ounces (1.7 kg). If delivered at this time, the baby has about a 50 percent chance for survival.

At the same time the fetus has usually assumed a head-down position in the mother's uterus. This position, called *cephalic presentation* (since the head appears at the cervix first during delivery), is the most common position for delivery and presents the fewest complications. During the seventh month, however, about 12 percent of fetuses are still upright in the mothers' uteruses (the so-called *breech presentation*), and a few are oriented with the long axis in a transverse position (called *shoulder presentation*). In addition to the baby's small size and immaturity, there is thus the added risk of a more complicated breech delivery early in the third trimester. At full term (nine months) only about 3 percent of babies are in the breech position.

By the end of the eighth month the fetus is about 18 inches (45.7 cm) long and weighs about 5 pounds, 4 ounces (2.4 kg). If delivered at this time, it has a 90–95 percent chance of survival.

During the ninth month the fetus gains more than 2 pounds (0.9 kg), and essential

[20]Wagner and Solberg, op cit.
[21]Goodlin et al. (1971). A more recent study (Wagner and Solberg, 1974) did not find any correlation between sexual activity and prematurity.

organs like the lungs reach a state of maturity compatible with life in the outside world. In addition, less crucial details like hair and fingernails assume a normal appearance and may even grow to such lengths as to require trimming shortly after delivery. At full term the average baby weighs 7.5 pounds (3.4 kg) and is 20 inches (50.8 cm) long.[22] Ninety-nine percent of full-term babies born alive in the United States survive, a figure that could be improved even further if all expectant mothers and newborn babies received proper medical care.

## Childbirth

As the end of pregnancy approaches, the expectant mother experiences contractions of her uterus at irregular intervals. The woman who has not delivered previously may rush to the hospital in the middle of the night only to be sent home because she is experiencing *false labor.*

Three or four weeks before delivery the fetus "drops" to a lower position in the abdomen. The next major step in preparation for delivery is the softening and dilation of the cervix. The mother may be unaware of this process, but usually just before labor begins there is a small, slightly bloody discharge (*bloody show*) that represents the plug of mucus that has been occluding the cervix. In about 10 percent of pregnant women, however, the membranes encapsulating the fetus burst (*premature rupture of the membranes*), and there is a gush of amniotic fluid down the woman's legs. Usually labor begins within 24 hours after such a rupture; but if it does not, there is risk of infection and the mother should be hospitalized for observation.

## Labor

The exact mechanism that initiates and sustains labor in humans is not fully understood, but a number of hormonal factors are known to be involved. Current research indicates that the fetus actually triggers labor.[23] The fetal adrenal gland is stimulated to produce hormones that act upon the placenta and uterus and cause these organs to increase the secretions of other chemicals, *prostaglandins.* Prostaglandins (*see* Chapter 6) stimulate the muscle of the uterus, thereby causing labor to begin. We have already mentioned the loss of the inhibitory effect on the uterus of placental hormones, particularly progesterone, as production drops off near term. Finally, *oxytocin,* a hormone produced by the posterior pituitary gland (*see* Figure 4.1) is released in the late stages of labor and stimulates the more powerful contractions required to finally expel the fetus.

Labor begins with regular uterine contractions ("pains"), which dilate the cervix. Labor is divided into three stages, the first of which is the longest, extending from the first regular contractions until the cervix is completely dilated—about 4 inches (10 cm) in diameter. This stage lasts about 15 hours in the first pregnancy and about 8 hours in later ones. (Deliveries after the first are generally easier in all respects.) Uterine contractions begin at intervals as far apart as 15 or 20 minutes, but they occur more frequently and with greater intensity and regularity as time passes. When the contractions are coming regularly four or five minutes apart the woman usually goes to the hospital, where she is admitted to a labor room (essentially a regular hospital room) for the remainder of the first stage. Her husband is usually allowed to remain with her during this time.

The second stage begins when the cervix is completely dilated and ends with delivery of the baby (*see* Figures 5.15, 5.16, and

---

[22]There is great variation in birth weights, however, ranging usually from five to nine pounds. Weights of ten or eleven pounds are not uncommon, and the largest baby known to have survived weighed 15.5 pounds at birth. Stillborn babies as heavy as twenty-five pounds have been delivered.

[23]Daly and Wilson (1978), p. 192.

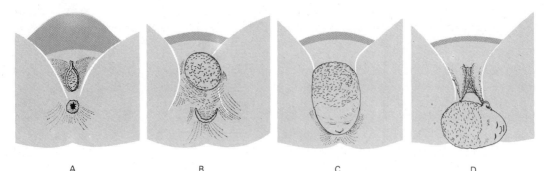

A        B        C        D

**Figure 5.15** (*A*) The scalp is visible at the vaginal opening. (*B*) More of the infant's head can be seen. (*C*) The head of the baby has emerged. (*D*) The shoulders of the baby are beginning to appear. The woman's pubic hair is absent in this figure. The pubic hair is often shaved when the woman enters the hospital. Redrawn from White, *Emergency Childbirth: A Manual* (Franklin Park, Ill.: Police Training Foundation, 1958) pp. 20–21.

5.17). The woman is taken to a delivery room (similar to a surgical operating room), and the husband may or may not be allowed to be present, depending on the laws of the state, hospital regulations, the discretion of the physician, and the wishes of the couple. The second stage may last from a few minutes to a few hours. If any anesthetic is used, it is usually given before this stage begins. General anesthesia for childbirth is becoming less popular than it was earlier in this century. Its disadvantages include all the risks to the mother that general anesthesia entails under any circumstances, plus a slowing effect on labor and depression of the infant's activity (particularly respiration). Caudal or spinal anesthetic is currently popular. The anesthetic is inserted

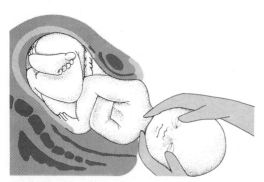

**Figure 5.16** Gentle traction to bring about descent of anterior shoulder. From Eastman and Hellman, *Williams Obstetrics*, 13th ed. (New York: Appleton-Century-Crofts, 1966), p. 423. Reprinted by permission.

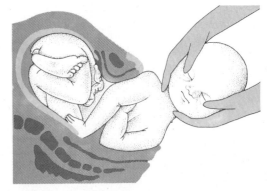

**Figure 5.17** Delivery of anterior shoulder completed; gentle traction to deliver posterior shoulder. From Eastman and Hellman, *Williams Obstetrics*, 13th ed. (New York: Appleton-Century-Crofts, 1966), p. 423. Reprinted by permission.

by needle into the spinal canal, producing temporary loss of sensation (and paralysis) below the waist only. Advantages include the mother's awareness through delivery and the infant's freedom from anesthetic effects.

In some cases local anesthesia, which simply blocks nerves in the vicinity of the vagina, is sufficient for a comfortable delivery in women who desire a minimum of medical intervention at this stage.

In the third stage the placenta separates from the uterine wall and is discharged as the *afterbirth* (placenta and fetal membranes). The uterus contracts to a significantly smaller size during this stage, and there is variable bleeding. The third stage of labor lasts about an hour, during which time the physician examines the mother and baby carefully and sews up any tear that may have occurred in the *perineum* (the skin and deeper tissues between the openings of the vagina and anus). The *episiotomy* incision, if there has been one, is also sewn up at this time. An episiotomy is an incision of the perineum that is sometimes performed to ease the passage of the baby's head. These "stitches" may cause itching and discomfort for several days, but they usually heal with no complications.

### Cesarean Section

Occasionally, as when the baby is too large (or the woman's pelvis is too small) to accommodate vaginal delivery, a cesarean section is performed. In this operation the baby is removed from the uterus through an incision in the abdominal and uterine walls. About 10 percent of deliveries in the United States are by cesarean section, although the figure is increasing and may be as high as 22 percent in some hospitals.

With modern surgical techniques the incidence of complications in this kind of delivery is no greater than that in vaginal delivery. It is not necessarily true that a woman who has had one cesarean section must have all her later children by this method: A cesarean delivery may be followed by normal vaginal deliveries. On the other hand, it is possible also to have several children by cesarean section, despite the common notion that a woman can endure only one such operation. The recovery period is somewhat longer after a cesarean section, and hospitalization usually lasts seven to ten days.

It is unlikely that Julius Caesar, for whom this operation was named, was actually delivered by cesarean section. Although it is known that the operation was performed in ancient times among both civilized and primitive peoples, it was almost always performed after the mother had died, and in hopes of saving the baby. Caesar's mother lived for many years after his birth. Cesarean section was definitely being performed on living women in the seventeenth century, but the operation at that time involved *removing* the uterus with the baby still inside and then sewing up the abdomen. The current practice of removing only the baby and leaving the uterus in place dates from 1882.

### Natural Childbirth

Intervention and assistance at childbirth in the form of rituals, prayers, potions, and medicines to relieve pain or dull the senses and physical extraction of the baby are as old as recorded history. In Europe childbirth usually occurred at home in familiar surroundings, and those assisting were women with special training and experience. The pattern changed significantly about 300 years ago. Male physicians and surgeons began to replace midwives; deliveries were moved from home to lying-in hospitals in large cities; and forceps (invented about 1600) became popular. While significant new knowledge in the field of obstetrics was acquired in the ensuing centuries, the clinical practice of obstetrics did not always improve the lot of women in labor. Many women died of childbed fever (*see* Box 5.4).

## Box 5.4   Childbed Fever

Before the mid-1800s, there was no knowledge that childbirth should take place in a clean, antiseptic environment. One woman in ten who gave birth in hospitals or clinics was likely to die of puerperal fever—childbed fever—and sometimes epidemics of puerperal fever took the lives of 50 to 75 percent of the women in maternity wards.

The maternity wards of early hospitals and clinics were often dark and poorly ventilated. Few people took precautions to prevent new mothers from being exposed to sources of infection. Oliver Wendell Holmes was among the first to realize in the 1840s that puerperal fever was infectious in nature; but his essay on the topic was denounced by colleagues and had little effect. Most doctors attributed puerperal fever to accident or Providence— the "curse of Eve"—and were certain that nothing could be done to alter the death rates.

In Vienna in 1847 a young doctor, Ignaz Semmelweiss, observed that the death rate of women who were attended by medical students was substantially higher than it was in wards where the women were attended by midwives. The patients were aware of this difference, too; and women begged to be assisted by mid-wives. Apparently, students often came directly to the maternity wards after handling the infected and the dead, thus passing the fever to the healthy women. Semmelweiss ordered the students to wash their hands in a solution of chlorinated lime before attending women in the maternity wards. The mortality rate from puerperal fever dropped from 18 percent to 3 percent within two weeks and to 1.27 percent with a year.

Semmelweiss's work was heavily criticized by his superiors. After months of persecution, he resigned his post and moved to another city. In 1861 he published his major work on puerperal fever, but the medical profession continued to reject this theory. He died in 1865 of puerperal fever contracted through a wound on his finger. A few years later, the work of Lister and Pasteur made the importance of germs and antisepsis better understood, and the incidence of puerperal fever began to decline permanently.

[1]Guthrie (1958)
[2]Gortvay and Zoltán (1968).

Chloroform anesthesia was introduced in the nineteenth century and became quite popular in England after Queen Victoria was delivered of her eighth child under chloroform in 1853. By the twentieth century childbirth had taken on the trappings of a surgical procedure, even when routine and uncomplicated. The woman was hospitalized, sterilized (in the antiseptic sense), anesthetized, and a variety of instruments were used in the delivery, which was performed in a room resembling a surgical suite in appearance, facilities, and regulations (for example, husbands and other "nonparticipants" were excluded).

It is not surprising that given these developments there would be a reaction to one or more aspects of the procedure. To some, the exclusion of the father was most repugnant; to others, the domination of obstetrics by men (in the United States) was objectionable; and for many, the most disliked aspect of twentieth-century childbirth was the use of instruments and anesthetics. Anesthetics have a depressant effect on the mother and infant. The widespread practice of performing episiotomies was also criticized as being unnecessary in many instances and even damaging in some cases.

The term "natural childbirth" was coined

by the English physician Grantly Dick-Read in 1932. Dick-Read postulated that the pain of childbirth is primarily related to fear and subsequent muscular tension during labor. Very briefly, his still-popular method of "natural childbirth" involves eliminating fear through education about the birth process before delivery in order to break the cycle of fear, tension, and pain. The method is best described in his book *Childbirth without Fear*.

A second type of "natural childbirth" originated in Russia, but was popularized by the French physician Bernard Lamaze. The "Lamaze method" is based on Pavlov's description of conditioned-reflex responses and is sometimes called the "psychoprophylactic method" of childbirth.[24] It involves conditioning the pregnant woman mentally to dissociate uterine contractions from the experience of pain through repeated reinforcement of the notion that such contractions are not painful. Prescribed exercises requiring voluntary relaxation of the abdominal muscles are also part of the program.

### Prepared Childbirth

The methods of *prepared childbirth* currently used in the United States incorporate a variety of techniques from Dick-Read, Lamaze, and other sources. The goal of prepared childbirth is to allow women to experience and to help in giving birth while feeling as little pain as possible. Although drugs are available if they are needed, the use of drugs is minimized to avoid risk of side effects to the mother and the baby. Usually, an expectant mother (and father) attend classes for 6 to 10 weeks before the birth in order to learn the birth procedures and practice techniques that will ease the birth process.

First, the classes are designed to inform prospective parents about each step of labor. This education helps answer the woman's questions and dispel her anxiety. Women

[24]Chertok (1967), pp. 698–699.

who learn the birth procedures in advance have less to fear and can help more in each stage of labor. In addition, they can be awake and aware of the remarkable experience of bringing a child into the world.

Second, the woman is taught a variety of exercises that increase her muscle control such that she can voluntarily tighten or relax specific muscles in various parts of her body. Part of the pain that women experience in childbirth results from their tightening their abdominal and perineal muscles and thereby making it harder for the baby to emerge. By learning to relax these and other muscles, women can allow the baby to pass through the birth canal more easily and thus decrease the total amount of pain. In preparation for the second stage of labor, the woman also learns the muscle contraction patterns needed to help push the baby through the birth canal while keeping the vagina and perineum relaxed.

Third, women are taught methods for distracting their attention from the experience of pain. They learn several breathing techniques that ease the birth process and keep their attention focused on systematic tasks. They learn patterns of hand movement—gently massaging the abdomen in a circular motion—that keep their hands busy, make them feel better, and help produce relaxation. They concentrate their visual attention on specified targets, and they hear their husbands counting and measuring the lengths of the contractions. These combined inputs from several sense modalities keep the woman's mind occupied with thoughts that help decrease the sensation of pain.

Finally, having the husband present during childbirth provides a source of comfort and security for the woman. The husband's reassuring words partially divert her attention from the pain. The husband learns to give gentle massages that bring comfort and help distract the woman's attention from other sensations. His concern and emotional support make the experience more positive for her. In addition, the husband makes sure that she continues her breathing exercises, keeps her abdominal muscles relaxed, and practices the other methods consistently throughout the birth process.

Although early advocates of "natural" childbirth claimed that women would feel no pain during birth, the present approach to prepared childbirth does not deny the existence of pain. Instead it helps women learn to cope with the experience in as positive a way as possible.

### Leboyer Method

Another method of childbirth that has become popular in recent years was developed by the French physician Frederick Leboyer. This method concentrates on protecting the infant from the pain and trauma of childbirth. In *Birth without Violence*, Leboyer describes what he believes to be the pain infants suffer being born and then goes on to describe a method of making birth less painful and shocking to the infant. It is a slow, quiet birth in which everything is done to protect the infant's delicate senses from shock. Lights are kept low, and unnecessary noises are avoided. When the infant's head appears the birth is eased along by the doctor's fingers under each of the infant's armpits. Supported so, the infant is gently settled onto the mother's abdomen. There, for several minutes, the child is allowed to adjust to its new environment while it continues to receive warmth and comfort from its mother as it did before birth. Instead of being held up and slapped (which helps the infant start breathing), the *Leboyer method* allows the child to lie quietly on its mother's abdomen with the umbilical cord still attached. The infant continues to receive oxygen from its mother in this way until it starts breathing on its own. Then the child is gently lowered into a warm bath where it will eventually open its eyes and begin to move its limbs freely. The result of such a

nonviolent birth, according to Leboyer, is not a screaming, kicking, terrified infant, but a relaxed and even smiling child.

If a woman can give birth without surgical intervention or anesthesia, one might wonder why the birth need take place in a hospital. The primary medical justification for hospital delivery is the availability of backup personnel and equipment in the event of a sudden complication (for example, hemorrhaging). However, there is a growing interest in the United States in "alternative birth centers" (clinics that are more homelike and family oriented and less expensive than hospitals), in home deliveries, and in midwives. One reason for this interest is the rising costs of hospitalization. Another reason is the impersonal atmosphere of hospitals and restrictions that separate husbands from wives during childbirth and mothers from their babies in maternity wards. Increasingly, hospitals have been modifying their regulations to allow husbands to be present throughout labor and delivery and to allow "rooming-in," that is, keeping newborn babies in the same room with their mothers rather than in a separate nursery. This allows the mother to hold and feed her new baby when she wishes, rather than according to a fixed schedule.

The Leboyer method and other so-called methods of natural childbirth are becoming increasingly popular in the United States with expectant couples as well as with some obstetricians and in some hospitals. But the term "natural" does not mean that the more traditional type of hospital delivery is wrong. Many women and children are alive today thanks only to the extraordinary hospital care they received and to the techniques of modern obstetrics.

## Multiple Births

The delivery of two or more infants after one pregnancy is an event of great fascination in all cultures. Twins occur in 1 of 90 births in the United States. Triplets occur in about 1 of 9000 births, and quadruplets in about 1 of 500,000 births. Multiple births of more than four children are extremely rare. The Dionne quintuplets of Canada, born May 28, 1934, were the first quintuplets known to survive. All five were girls; and they weighed a total of 13 pounds, 6 ounces at birth. Because of their small sizes and usually premature delivery, the mortality rate among infants in a multiple birth is significantly higher than with single births. Twins are born an average of 22 days before the EDC. Their mortality rate is two to three times that of single births.

There are two types of twins, identical and fraternal. Two out of three sets of twins are fraternal, developed from two separate eggs fertilized simultaneously. It is biologically possible for twins to have different fathers. This phenomenon is called *superfecundation* and has been clearly documented in animal studies. In 1975 a German woman gave birth to fraternal twins—one fathered by a black man and the other fathered by a white man. She had had intercourse with both men at different times on the same day.

Another rare occurrence is *superfetation,* the fertilization and subsequent development of an egg when a fetus is already present in the uterus. There are a few such cases reported in the medical literature. In one instance, a woman gave birth to two normal children three months apart.

Identical twins result from subdivision of a single fertilized egg before implantation. Identical twins usually share a common placenta, whereas fraternal twins usually have separate placentas.

Twins show a slight tendency to ''run in families,'' but the heredity of ''twinning'' is rather vague. Fraternal twins are more likely to reoccur than are identical twins.

## The Postpartum Period

A woman nowadays usually leaves the hospital two or three days after an uncomplicated delivery, though the length of the stay varies somewhat from place to place. With a first baby the first week at home may be a bit of a turmoil for the young mother, who is trying to cope simultaneously with the needs of the baby (feeding, bathing, changing diapers) and the calls of many well-meaning friends and relatives anxious to see ''the new addition.'' Fatigue may be a major complaint at this point, along with a general ''let-down'' feeling.

About two-thirds of all women experience transient episodes of sadness and crying some time during the first ten days after delivery, a phenomenon known as the ''postpartum blues'' syndrome. Women (and their families) are often puzzled by this reaction, for it comes when they are expected to be especially happy, celebrating the arrival of the new child. Doubts about their competence as mothers, fatigue, feelings of rejection or neglect by their husbands, and the drastic hormone changes that occur at this time are some factors involved.[25]

## Breast-Feeding

Milk production (*lactation*) begins 48 to 72 hours after childbirth and is accompanied by a feeling of engorgement in the breasts. Two pituitary hormones are involved in the physiology of breast-feeding. *Prolactin* (*see* Chapter 4) stimulates the production of milk by the mammary glands, and *oxytocin* causes the ejection of the milk from the breast to the nipple in response to the stimulus of suckling by the baby. At the time of weaning, the breast is no longer subject to the stimulation of the nursing infant, with the result that secretion of prolactin and oxytocin, and therefore lactation, ceases.

Breast-feeding is currently becoming more popular in the United States than in earlier times, and organizations such as La Leche League provide information and encouragement to women who are interested

[25]Yalom et al. (1968).

## Box 5.5    Mammaries

Mammals include all animals that feed their young with milk from the female mammary glands. Many mammals bear a litter of young at each birth and have several mammaries to provide milk for them. Although humans normally have only two mammaries, about 1 percent of both males and females are born with extra nipples and mammaries.[1] The extra mammaries can develop at any location along the milk lines, which extend from the armpits to the inguinal regions. Not all nipples have mammary tissue, but those with mammary tissue are capable of lactation. The presence of such extra nipples reveals the similarities in body plan between humans and the other mammals. The extra mammaries can be removed surgically without complications.

The male mammary is potentially capable of the full development seen in adult women. Transsexuals who were born male and take female hormones develop breasts with the typical female structure and shape. The male breast has the potential to produce milk. When both male and

female babies are born, they may have droplets of milklike substance (known as "witch's milk") in the nipples. Before birth, the fetus's breasts had been stimulated by the mother's hormones in the uterus, hence the fetus's breasts—whether male or female—began to develop the capacity to produce milk much the same as the mother's breasts did.

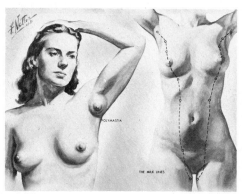

From Netter, *Reproductive System*. The Ciba Collection of Medical Illustrations, vol. 2. © Copyright 1954, 1965 CIBA Pharmaceutical Company, Division of CIBA-GEIGY Corporation. Reproduced with permission. All rights reserved.

[1]Netter (1965).

in learning more about it. The experience of nursing, from the mother's standpoint, can be emotionally satisfying, pleasurable, and even sensual. In fact, there is a positive correlation between postpartum sexual interest and nursing. There may be some discomfort at the beginning, but this can be overcome with patience. Conversely, women with an aversion to nudity and sexuality are less likely to breast-feed.[26]

From the standpoint of the baby there is no question of the superiority of human milk over cow's milk or commercial formulas. Human milk contains the ideal mixture of nutrients for human babies; it contains antibod-

ies that protect the infant from certain infectious diseases; it is free of bacteria; and it is always the right temperature.

In many countries cow's milk is neither hygienic nor cheap; and the water available for dilution of formula preparations may be contaminated. Nevertheless, the trend in many areas of the world has been to abandon the breast for the bottle. The results could be nutritionally and economically disastrous if this trend continues.

## Resumption of Ovulation and Menstruation

There is wide variation in timing of the resumption of normal menstrual cycles after delivery. For about four weeks the woman

[26]Newton (1973).

has a bloody vaginal discharge called *lochia* (from Greek *lokhios,* "pertaining to childbirth"). One or two months later a mother who is not nursing may have a menstrual period. If she is nursing, however, her periods may not resume for as long as 18 months, although 5 months is more common. The first few periods after pregnancy may be somewhat irregular in length and flow, but women who have had painful periods before pregnancy often find that they suffer no such discomfort after childbirth.

It should be emphasized that ovulation can occur *before* the first postpartum menstrual cycle and that consequently a woman can become pregnant without having a menstrual period after the birth of her baby. It should also be noted that women can conceive while nursing. The notion that nursing is "nature's method of contraception" is misleading.

## Sexual Activity during the Postpartum Period

There is considerable variation in patterns of sexual activity after delivery. Fatigue, physical discomfort, and the obstetrician's injunctions play an important part in determining when a woman resumes sexual relations after childbirth. Although doctors commonly advise women to refrain from intercourse for six weeks after delivery, there is no medical reason why a healthy woman cannot have vaginal intercourse as soon as the episiotomy or any lacerations of the perineum have healed and the flow of lochia has ended. The only medical concern is the possibility of infection through the vagina. Couples who practice intracrural intercourse, manual manipulation of the genitals to orgasm, or sexual activity other than vaginal intercourse early in the postpartum period are not hampered by this concern.

Studies have shown that a certain percentage of men feel driven to extramarital sexual affairs during the period of absti-

nence—"six weeks before and six weeks after"—that is rigidly adhered to by some women. That this explanation is a rationalization is certainly possible, because the imaginative couple that considers regular sexual activity desirable can find satisfying forms of gratification during periods when vaginal intercourse is inadvisable.

## Pregnancy and Childbirth in Other Societies

Preliterate people have developed various beliefs and practices in connection with pregnancy and childbirth. Many tribes believed that the fetus developed from a combination of male semen and menstrual blood. The Venda tribe of East Africa believed that "red elements" like muscle and blood were derived from the mother's menses (which ceased during pregnancy because the menstrual blood was being absorbed by the developing fetus). The "white elements"—like skin, bone, and nerves—were believed to develop from the father's semen.

Some societies imposed dietary restrictions on pregnant women, often from fear that the fetus might otherwise take on undesirable characteristics of food, plants, or animals. For example, if the mother ate rabbit, the child might have weak legs; if she ate trout, he might exhibit characteristic quivering movements. In addition, Ashanti women were forbidden to look upon any deformed object or creature during pregnancy lest their children be born with similar deformities.

The majority of primitive tribes that have been studied prohibited sexual intercourse during the last month of pregnancy on the grounds that it might kill the child or cause premature delivery, an interesting observation considering the similar concern in modern societies and the absence so far of substantial medical evidence for or against this belief.

Abortions were performed in some societies, particularly if the women were un-

**Box 5.6**   Natural Childbirth in Other Societies

Natural childbirth is quite different from the prepared childbirth common in the United States today. Women have been giving birth the natural way all through history, and many have suffered more than the woman does in modern prepared childbirth. Women in preliterate societies have practiced many methods of natural childbirth.[1,2] Sitting, squatting or kneeling positions were very common. In addition, various forms of assistance have been given to "facilitate" birth. The following methods are among those that have been reported: suspending the woman from a limb; massaging the abdomen; sitting or standing on the abdomen; blowing smoke into the vagina; and tossing the woman with a blanket to shake the baby out of her body.

[1]Englemann (1883).
[2]Witkowski (1892).

married or were pregnant as the result of adultery. Usually the fetuses were killed *in utero* by violent beating upon the abdomen, and this was followed by mechanical extraction of the fetus or spontaneous stillbirth.

Contrary to popular opinion, childbirth was not considered a routine and painless event by most preliterate people. Among many tribes elaborate dietary and exercise regimens were practiced to prevent painful and difficult deliveries. Various rituals might be performed to ensure easy delivery, and particular attention was paid to confession of sexual indiscretions at this time, for difficult deliveries were often attributed to violations of the tribal sexual codes.

The placenta was almost always viewed as a potentially dangerous object and was carefully disposed of, usually by burying it in a special place. There was also usually a taboo on sexual intercourse for several weeks or more after delivery, another striking similarity to taboos in modern societies.

Deformed babies and multiple births were viewed with alarm in most primitive societies. Twins, triplets, and babies with congenital deformities were usually killed at birth. Twins were often believed to result from adultery or impregnation by an evil spirit. A more benign explanation was offered by the Kiwai tribesmen of New Guinea, who believed that a woman would give birth to twins if she ate bananas from a tree with two bunches!

# Chapter 5

# Contraception

Some of the oldest documents to mention contraception (voluntary prevention of conception) date back nearly 4000 years to ancient Egypt.[1] One method called for douching, or washing, the vagina after intercourse with a mixture of wine, garlic, and fennel. The practice of coitus interruptus is mentioned in *Genesis* 38:8–9, as well as in the Talmud. Absorbent materials, root and herb potions, pessaries, and more permanent means of sterilization were used by the Greeks and Romans and have been used subsequently by many peoples around the world.

Prior to the present era it was very rare for any society to limit population as a matter of general policy. A few primitive cultures practiced selective infanticide, usually of female infants; and some peoples living in unusually harsh environments, such as the Eskimos, allowed the very old to die of starvation. However, the historical incentives for birth control were usually personal or idiosyncratic: Slaves, harlots, and illicit lovers tried to avoid pregnancy; and other people sought to regulate the sizes of their families because of economic, medical, or psychological factors. Such personal incentives continue to be important, but there is now a mounting concern that entire nations and even humanity as a whole must urgently check galloping population expansion.

In the United States contraception has had a turbulent history. In 1873 Congress passed the Comstock law, which made it illegal to disseminate contraceptive information. Soon state laws were also passed, some of which were even more restrictive than the federal law.

Margaret Sanger was a leader in the birth control movement. She became aware of the need for contraception during her work as a nurse in the poor tenements of New York City in the early 1900s. There she saw the misery and suffering caused by a lack of contraceptive knowledge. After one particularly difficult experience (*see* Box 6.1), she vowed "to do something to change the destiny of mothers whose miseries were as vast as the sky."[2] She devoted the rest of her life to the cause of helping women gain control over their own bodies. In the face of strong opposition she worked to make birth control available to all women. Her early publications were banned and she was indicted on nine federal offenses. In 1916 she opened the first birth control clinic in the United States, only to be arrested and have the clinic closed. When released on bail she reopened the clinic; this time she was arrested and sentenced to 30 days. But she was having her impact. In 1918 the courts ruled that doctors could give contraceptive information for pre-

---

[1]Suitters (1967).

[2]Sanger (1938).

## Box 6.1   Margaret Sanger

Then one stifling mid-July day of 1912 I was summoned to a Grand Street tenement. My patient was a small, slight Russian Jewess, about twenty-eight years old, of the special cast of feature to which suffering lends a madonna-like expression. The cramped three-room apartment was in a sorry state of turmoil. Jake Sachs, a truck driver scarcely older than his wife, had come home to find the three children crying and her unconscious from the effects of a self-induced abortion. He had called the nearest doctor, who in turn had sent for me. Jake's earnings were trifling, and most of them had gone to keep the none-too-strong children clean and properly fed. But his wife's ingenuity had helped them to save a little, and this he was glad to spend on a nurse rather than have her go to a hospital.

The doctor and I settled ourselves to the task of fighting the septicemia. Never had I worked so fast, never so concentratedly. The sultry days and nights were melted into a torpid inferno. It did not seem possible there could be such heat, and every bit of food, ice, and drugs had to be carried up three flights of stairs.

Jake was more kind and thoughtful than many of the husbands I had encountered. He loved his children, and had always helped his wife wash and dress them. He had brought water up and carried garbage down before he left in the morning, and did as much as he could for me while he anxiously watched her progress.

After a fortnight Mrs. Sachs' recovery was in sight. Neighbors, ordinarily fatalistic as to the results of abortion, were genuinely pleased that she had survived. She smiled wanly at all who came to see her and thanked them gently, but she could not respond to their hearty congratulations. She appeared to be more despondent and anxious than she should have been, and spent too much time in meditation.

At the end of three weeks, as I was preparing to leave the fragile patient to take up her difficult life once more, she finally voiced her fears, "Another baby will finish me, I suppose?"

"It's too early to talk about that," I temporized.

But when the doctor came to make his last call, I drew him aside. "Mrs. Sachs is terribly worried about having another baby."

"She well may be," replied the doctor, and then he stood before her and said, "Any more such capers, young woman, and there'll be no need to send for me."

Margaret Sanger The Sophia Smith Collection, Smith College, Northampton, Mass.
Reprinted from *Margaret Sanger: An Autobiography* (New York: Norton, 1938). By permission.

**Box 6.1    Continued**

"I know, doctor," she replied timidly, "but," and she hesitated as though it took all her courage to say it, "what can I do to prevent it?"

The doctor was a kindly man, and he had worked hard to save her, but such incidents had become so familiar to him that he had long since lost whatever delicacy he might once have had. He laughed good-naturedly. "You want to have your cake and eat it too, do you? Well, it can't be done."

Then picking up his hat and bag to depart he said, "Tell Jake to sleep on the roof."

I glanced quickly at Mrs. Sachs. Even through my sudden tears I could see stamped on her face an expression of absolute despair. We simply looked at each other, saying no word until the door had closed behind the doctor. Then she lifted her thin, blue-veined hands and clasped them beseechingly. "He can't understand. He's only a man. But you do, don't you? Please tell me the secret, and I'll never breathe it to a soul. *Please!* "

What was I to do? I could not speak the conventionally comforting phrases which would be of no comfort. Instead, I made her as physically easy as I could and promised to come back in a few days to talk with her again. A little later, when she slept, I tiptoed away.

Night after night the wistful image of Mrs. Sachs appeared before me. I made all sorts of excuses to myself for not going back. I was busy on other cases; I really did not know what to say to her or how to convince her of my own ignorance; I was helpless to avert such monstrous atrocities. Time rolled by and I did nothing.

The telephone rang one evening three months later, and Jake Sachs' agitated voice begged me to come at once; his wife was sick again and from the same cause. For a wild moment I thought of sending someone else, but actually, of course, I hurried into my uniform, caught up my bag, and started out. All the way I longed for a subway wreck, an explosion, anything to keep me from having to enter that home again. But nothing happened, even to delay me. I turned into the dingy doorway and climbed the familiar stairs once more. The children were there, young little things.

Mrs. Sachs was in a coma and died within ten minutes. I folded her still hands across her breast, remembering how they had pleaded with me, begging so humbly for the knowledge which was her right. I drew a sheet over her pallid face. Jake was sobbing, running his hands through his hair and pulling it out like an insane person. Over and over again he wailed, "My God! My God! My God!"

---

venting and curing disease. This was the beginning of many changes.

Contraception remains a controversial issue. This is not surprising when one considers the profound emotions that have come to be associated with reproduction as the result of the long history of our attempt to survive as a species. Indeed, the fact that reliable birth control has not been available and practiced throughout most of human history accounts in part for our survival. But today, the survival of *Homo sapiens* may depend on a reversal of attitudes and a change in birth control practices. Although opposition to birth control programs is gradually diminishing, it is still formidable in certain parts of the world and in certain segments of our own society. Among married people in the United States who are at risk of an unplanned pregnancy, nine out of ten couples use some form of contraception.[3]

[3]Tietze 1979(a).

## Preventing Unwanted Pregnancies

The overwhelming majority of people who currently use contraceptives do so for personal and private reasons: avoiding pregnancy out of wedlock; postponing pregnancy for economic and psychological reasons (as when the future of a marriage is uncertain); preventing the birth of a deformed or seriously ill child; and limiting family size because of economic, health, and other considerations.

Moral judgments in these matters vary widely. Although some people do not use contraceptives under any circumstances, most people approve of their use for certain purposes—"Their use is legitimate if a woman's life will be otherwise endangered by pregnancy"; "Married people may use them"; "People about to be married may use them"; "All adults may use them."

The use of contraceptives is becoming progressively more legitimate, despite opposition. If moral considerations were not involved, the realistic approach would be to prescribe contraceptives to the entire postpubertal population upon demand. The common fear is that dispensing information and contraceptive devices tacitly encourages youngsters to engage in intercourse. The large number of adolescent pregnancies indicates however, that lack of contraceptive knowledge or of the willingness to use contraception is not discouraging many teenagers from engaging in sex.

There are, on the other hand, countless educated and financially secure adults who become pregnant only unintentionally or for questionable reasons: to perpetuate family names, to please the grandparents, to bolster failing relationships, to "tie a partner down," to force partners into marriage, to conform socially, and so on.

The true motivations for having children are often rationalized or unconscious. Parents may view children as extensions of themselves. Through their children parents may seek to gratify their own unfulfilled childhood needs, to replay early dramas and this time to come out the "winner." Parents may regard their offspring as a means to fill their own inner void, to give meaning to their lives, and to consolidate a faltering sense of self-worth. Pregnancy may result from conscious or unconscious hostility, as an aggressive or punitive act against the sexual partner who does not want parenthood.

A surprising number of sophisticated unmarried young men and women risk unwanted pregnancies because taking contraceptive measures implies forethought. They feel that such precautions rob sex of its spontaneity and turn it into a calculated and unemotional activity. Such "refined" considerations, however, frequently conceal confusion and guilt about premarital sex. Consider, for instance, a couple in college. They are fond of each other and would like to have intercourse, but one or the other feels that they should not. They engage in progressively heavier petting, which ultimately does culminate in intercourse without contraceptives. Although stricken with remorse, they try to alleviate their guilt by claiming that they did not plan intercourse, that they could not help it. (Occasionally alcohol is used as the scapegoat in these encounters.) Premarital sex has definite moral implications that must be faced. A person may decide for or against it for a variety of reasons, but the difficult decision-making process cannot be evaded without risks to both parties involved.

Until very recently, little was known about contraceptive behavior among those who are sexually active but wish to avoid pregnancy. Sufficient data are now available to support these observations.[4] Among sex-

---

[4]For further information on current contraceptive behavior in the U.S. among various age, socioeconomic, ethnic, and religious groups, see Kantner and Zelnick (1972, 1973); Miller (1973); Shah, Zelnik and Kantner (1975); Westoff and Bumpass (1973); Zelnick and Kantner (1977); and Zelnik, Kim, and Kantner (1979).

ually active teenagers, birth control pills, condoms, and withdrawal are the most common methods; but about 37 percent report using no method at last intercourse. Also, about one-fourth of sexually active teenagers say they *never* use contraception. The pregnancy rates for teens who are sexually active before marriage reflect their failure to use contraception. Thirty-five percent of these women become pregnant before they are 19. Among unmarried college students the same methods are most popular. At least 10 percent of unmarried sexually active college students report *never* using contraceptives and, not surprisingly, one recent study of college seniors found that 9.5 percent of the women had become pregnant (and had abortions) during their first 3½ years of college.

Concerning married adults a particularly striking change in contraceptive behavior has occurred among American Catholics. The proportion of Catholic women using a contraceptive method other than rhythm increased from 30 percent in 1955 to 51 percent in 1965, and between 1965 and 1970 increased to 68 percent. Of the Catholic women who were married between 1970 and 1975, 90.5 percent used methods other than rhythm.[5]

There is now a great need for safer, more reliable means of contraception, a particularly poignant need, considering that approximately one-third of the births to married women in the United States are unintended—either unwanted or mistimed.[6] This number is even larger for teenage pregnancies. A survey of teens who had had unprotected intercourse revealed that only 7 percent wanted to become pregnant and an additional 9 percent did not mind if they became pregnant.[7] One study, based on interviews of all pregnant women who came to a university medical center obstetrics clinic during a one-month period, revealed that only one-fourth of the cases involved planned pregnancies in which both parents looked forward to the baby's arrival. The remaining three-fourths fell into three groups of about equal size, according to whether their "rejection" was rated severe, moderate, or mild.

A woman considered to show "severe rejection of pregnancy" would be extremely agitated and upset about being pregnant and would state unequivocally her objections to having a child. She would generally either have practiced a "reliable" contraceptive method that had failed or would volunteer the information that she had attempted to have an illegal abortion or would have had an abortion had it been legal. (This study was conducted in California one year before the abortion law was liberalized.)

Women in the category of "moderate rejection of pregnancy" expressed ambivalence about their pregnancies, but gave very firm reasons for not wanting children. They also had generally been using contraceptive methods that had failed them.

Women in the category of "mild rejection of pregnancy" were unhappy about being pregnant at that particular time, though they declared previous plans to have children at some later date. The women were often using methods of contraception that they knew to be somewhat unreliable, such as the rhythm method. In summary, although varying degrees of regret over pregnancy were revealed in this study, only 25 percent of the sample definitely wanted the children that they were about to deliver.[8]

Even with legalized abortion, many women give birth to unwanted children because abortions are not readily available to them or because they find abortion personally unacceptable.

[5]Westoff and Jones (1977).
[6]Tietze (1979a).
[7]Shah, Zelnick, Kantner (1975).

[8]Yalom et al. (1968), p. 18.

## Contraceptive Methods

The effectiveness of contraceptive methods is measured in terms of *failure rates.* For example, the condom has a theoretical failure rate of 3 percent, which means that three women out of a hundred would become pregnant in one year if the condom were the only method of contraception used during that year. Because people do not always use contraceptives as carefully as they should, actual failure rates are higher than the theoretical rates. The average failure rate associated with the condom in actual practice is 10 percent, which means that ten women out of a hundred would become pregnant in one year if the condom were the only method of contraception used during that year. The theoretical and actual failure rates for all the contraceptive methods are summarized in Table 6.2 at the end of the chapter.

## The Pill

No drug since penicillin has been as rapidly and as widely accepted as *the pill,* the popular name for a number of commonly used oral contraceptives. The pill was first put on the market in 1960, and by 1977, 54 million women around the world were using it.

### History

Early in this century researchers discovered that ovulation does not occur during the luteal, or postovulatory, phase of the menstrual cycle (*see* Chapter 4) or during pregnancy. It was also discovered that those are the times at which the levels of progesterone are highest. The connection seemed obvious—progesterone prevents ovulation. This conclusion was tested after progesterone was isolated and purified in the laboratory in 1934. When administered to rabbits, progesterone was shown to inhibit ovulation and to prevent pregnancy. Estrogen, another female sex hormone, was chemically isolated at about the same time, and by 1940 it was being used to treat certain menstrual disorders.

The next step in the development of an oral contraceptive was taken in 1954 when Carl Djerassi succeeded in synthesizing in the laboratory a group of steroid chemicals called *progestogens.* The term comes from the word "gestation" and refers to the ability of these synthetic compounds to bring about "pseudopregnancy," or fake pregnancy. The progestogens produce certain changes in the lining of the uterus (endometrium) and elsewhere in the reproductive system that are similar to changes seen during pregnancy. Once these changes have taken place a woman usually will not ovulate and cannot get pregnant. The synthetic compounds were found to be much stronger than natural hormones and could therefore be used in much smaller doses to prevent pregnancy.

Margaret Sanger and Katharine McCormack supported and financed the work of Gregory Pincus (sometimes called the "father of the pill") and coworkers J. Rock and C.R. Garcia in the search for a better method of contraception. This team of researchers began testing about 200 of these new compounds on animals. Several were found to be particularly effective as antifertility agents, and the first large-scale field trials on human beings of a contraceptive pill were initiated by Pincus and his colleagues in San Juan, Puerto Rico, in 1956. The drug was highly successful in these initial trials, and within a few years the era of "the Pill" had clearly begun.

### Effects

Most contraceptive pills contain synthetic compounds resembling progesterone and estrogen. Studies of animals and humans have shown that progestogens tend to inhibit secretion of LH by the pituitary, while synthetic estrogens inhibit secretion of FSH. Both LH and FSH are essential to ovulation (*see* Chapter 4), and the pill is believed to work primarily by preventing ovulation. The only sure way to find out if this is the case is by microscopic examination of the ovaries and fallopian tubes of a woman who has been taking the pill.[9] Such observations have been made during hysterectomies (surgical removal of the uterus), and so far no eggs have been found in the tubes of women using the pill. It is still possible, however, that the woman may have ovulated earlier in her period or that she would have done so later.

Even if the pill does not prevent ovulation, it brings about other changes that can prevent pregnancy. There is some evidence, for instance, that the pill has a direct effect on the ovarian follicles and prevents maturation of the ova. It is also possible that the pill increases the rate at which the egg travels to the uterus, causing it to arrive there before it is sufficiently mature or before the lining of the uterus is ready to receive it.

A delicate balance of progesterone and estrogen is necessary if the lining of the uterus is to be receptive to implantation of the egg. However, changes in the endometrium have been seen in women taking the pill, and these changes may contribute to temporary infertility. In addition, the cervical mucus is changed by the pill (it becomes thicker and has a more acidic pH), and this may act as a barrier to the sperm on their way to the uterus.

### Usage

The contraceptive pills commonly available (with a physician's prescription) contain a mixture of synthetic progestogens and estrogen compounds and thus are called *combination pills.* Each pill contains from 0.3 to 10.0 milligrams of the progestogen (depending on the specific compound used and on the manufacturer) and a much smaller amount of estrogen, from 0.02 to 0.15 mg per pill. These pills are usually taken for 20 or 21 days

of the cycle, beginning on the fifth day after the start of menstruation. If the pills are not begun until the sixth day, they are still effective, but if they are not started by the seventh or eighth day, there is a risk of failure.

After three weeks the pills are stopped. "Withdrawal bleeding" (considered by some not to be "true" menstruation because ovulation presumably has not occurred) usually begins three to four days after pill taking has stopped. The first day of withdrawal bleeding is considered the first day of the next cycle, and the pills are resumed on the fifth day thereafter. Common brand names for the pill are *Enovid, Ortho-Novum, Ovulen,* and *Norinyl.*

Some manufacturers now produce and recommend a 21:7 pill program. The pills are the same as before, but the woman takes 21 pills, then stops for seven days, then repeats the series again, regardless of when menstruation begins. The main advantage of this regimen is that the woman always starts taking her pills on the same day of the week. Another way to help a woman to remember to take the pill is to have her take one every day—21 hormone pills followed by seven placebo (inert) pills packaged in such a way as to prevent her from taking the wrong pill on any given day.[10]

Many new contraceptive pills are coming onto the market, but they differ from older types mainly in the specific proportions of the progestogen and estrogen compounds. The trend has been toward smaller doses of both of these compounds in order to reduce side effects. Whereas most early birth control pills contained 10.0 mg of progestogen, later this amount was reduced to 5.0 mg and then to

[9]There are indirect indications like elevation of the basal body temperature (BBT) and excretion of a progesterone metabolite called "pregnanediol."

[10]As there is relatively little difference in composition among most birth control pills, competitors have tried to improve packaging as a selling point. Some pills come in fancy plastic cases rather than in paper containers. Some packages have built-in calendars or other "reminders," and some have dispensers indicating which pill is to be taken each day by date, which eliminates the worrisome question: "Did I take my pill today?"

2.5 or 1.0 mg by several manufacturers. The progestogen content is as low as 0.30 to 0.50 mg in some pills. The estrogen content was reduced from the early 0.10–0.15 mg to 0.02–0.05 mg per pill. Until 1976 a "sequential" pill was also marketed. These pills were taken on a 20-day cycle, but the first 15 pills contained only estrogen; the remaining five contained estrogen and progesterone. This was supposed to be more like the normal sequence of hormonal events during the menstrual cycle, but the sequential pills had several serious drawbacks. They were less effective and were possibly associated with a risk of blood clotting and cancer. When this was discovered, the sequential pills were removed from the market.

### The Mini-Pill

In 1973 the first so-called *mini-pill* was marketed, containing only 0.35 mg of progestogen and no estrogen. The mini-pill is slightly less effective than the combination pill, with a theoretical failure of 1.5 percent. The actual failure is closer to 5 or 10 percent. The irregular menstrual cycles and spotting associated with the mini-pill make it unacceptable to some women.

### The "Morning-After" Pill

The "morning-after" pill was approved by the Food and Drug Administration (FDA) in 1973 for use in emergency situations only (rape, incest, or where, in the physician's judgment, the woman's physical or mental well-being is in jeopardy). It contains a potent estrogen, diethylstilbestrol (DES), taken in a dosage of 25 mg twice a day for five days. To be effective in preventing pregnancy treatment must begin within 72 hours of unprotected intercourse, preferably within 24 hours. The most common side effect, occurring in about 16 percent of women, is nausea and vomiting.[11] The drug may also cause cancer in female offspring of women who take the

[11]Kuchera (1972), p. 177.

drug while pregnant (*see* Chapter 7). The *FDA Bulletin* (May 1973) states:

There is at present no positive evidence that the restricted postcoital use of DES carries a significant carcinogenic risk either to the mother or the fetus. However, because existing data support the possibility of delayed appearance of carcinoma in females whose mothers have been given DES later in pregnancy, and because teratogenic and other adverse effects on the fetus with the very early administration recommended are ill understood, failure of postcoital treatment with DES deserves serious consideration of voluntary termination of pregnancy.

### Effectiveness of the Pill

There is no question that birth control pills, when properly used, are the most effective contraceptive measure available today except for surgical sterilization. Taken as directed they are virtually 100 percent effective in preventing pregnancy by the second month of usage. Effectiveness is slightly less than 100 percent during the first month. If a woman does become pregnant while using the pill, it is probably because she failed to take it regularly. Forgetting one pill is not usually significant, provided that the woman takes two the next day, but there is a fair risk of failure if pills are skipped for two days or more. In order to minimize the risk of pregnancy in such cases, a woman who has missed two or more pills should rely on another form of contraception (such as condoms) for the remainder of the cycle. The actual failure rate of about 4 to 10 percent reflects the fact that women do make mistakes in taking their pills.

### Side Effects[12]

The users of oral contraceptives are probably one of the largest and most closely-watched groups in the history of medical sci-

[12]For a more comprehensive review of this topic, see Rinehart and Piotrow (1979), Potts et al., (1975), and Rudel et al. (1973), chapter 3.

ence. The reason for such close scrutiny is the large number of reports of side effects and possible side effects. The most common complaints of users of the pill are *nausea, weight gain, headaches,* and *vaginal discharge.*

Nausea is an estrogen-dependent side effect, and pills with smaller amounts of estrogen (or none at all) may be recommended if the symptom persists. In many cases nausea and other side effects diminish and then disappear after two or three months of usage.

Weight gains occur in 5 to 25 percent of women, depending on the particular preparations used. Partly these gains result from greater fluid retention caused by progestogens. They can be countered by drugs (diuretics) or sometimes by switching to different birth control preparations. Most weight gains, however, are caused by actual accumulation of fat, especially in the thighs and breasts. They occur mainly in the first month of using the pill and are partly related to increased appetite, not unlike the increased appetite that occurs in pregnancy. The additional pounds gained in the first few months often remain as long as the woman is on the pill, however. Weight gains are most common with high-dosage combination pills.

It is not clear why headaches are more frequent among pill users; but there may be a connection with some of the documented physiological side effects of the pill, which include increased blood pressure, fluid retention, and alteration of thyroid, adrenal, and blood sugar control mechanisms.

There is substantial evidence for an increase in vaginal discharge and vaginal infections with the use of oral contraceptives. The pill alters the chemical composition of the cervical secretions and the composition of the vaginal flora (microorganisms that normally inhabit the vagina and are not pathogenic). These factors make the vagina more susceptible to fungus infections (*see* Chapter 7). Unlike nausea and other gastric complaints, this side effect is more likely to be related to prolonged progesterone therapy.

Breakthrough bleeding, or minor "spotting," while taking birth control pills is annoying but can often be remedied by switching to a different preparation.

Acne sometimes improves with oral contraceptives, but pigmentation of the face (*chloasma,* the so-called "mask of pregnancy") and eczema can be undesirable skin manifestations of prolonged pill usage.

Less common side effects include tenderness of the breasts, nervousness, depression, alterations in libido patterns, menstrual irregularities, and general malaise. Women with histories of liver disease (for example, hepatitis) may find the ailments reactivated by the pill, and they must take it only under strict medical supervision.

### Subsequent Fertility

Contraceptive pills do not change subsequent fertility, although a few women may experience a delay in returning to normal fertility levels.[13] There is no evidence that prolonged use of the pills either hastens or postpones menopause.

Although the pill is known to be responsible for certain side effects, there are indications that some of the side effects blamed on the pill may be due to psychological rather than to physiological factors. In one study, for instance, the frequency of such complaints as dizziness, headache, nervousness, and depression was essentially the same for IUD users as for oral contraceptive users. In another study, women who thought they were receiving oral contraceptives but who were actually receiving placebos (inactive pills) reported a higher incidence of side effects, including decreased sexual drive. And in a third study of women using oral contraceptives, the researchers changed the color of the pills every six months and found a change in libido with each new pill color. Libido would gradually return to the previous level only to drop again with each color change.

[13]Potts et al. (1975); Vessey (1978).

Certain beneficial side effects of the pill have also been found and have led to their prescription even when contraception is not the major goal. For example, the pill can help relieve premenstrual tension and eliminate menstrual irregularity and pain, as well as other menstrual disorders. Pill use has also been linked with decreased incidence of ovarian cysts, rheumatoid arthritis, and iron deficiency anemia. In addition, some women report a general sense of well-being, as well as increased pleasure in sexual intercourse, while using the pill. This last effect is probably partially due to elimination of the fear of unwanted pregnancy.

### When Not To Use the Pill

Alerting people to the possible problems associated with pill usage can decrease the incidence and severity of some pill-related side effects. Recent research has shown that the pill can have serious side effects that, while rare, can be fatal. Large-scale studies clearly indicate that there is a small, though statistically significant, increase in the incidence of blood clotting with subsequent complications in women who take the pill. A blood clot (*thrombus*) may, for instance, form in one of the deep veins of the leg in association with a local inflammation (*thrombophlebitis*). The clot may then break loose (as an *embolus*) and travel toward the lungs, where it may block a major blood vessel (as a *pulmonary embolus*). The result can be fatal. There is also a slightly greater chance of a blood clot in the brain (*thrombotic stroke*) among users of the pill. The risk of death from blood clots is between 1.5 and 4 per 100,000 users versus 0.4 per 100,000 among nonusers.[14] The risks are greater than this for women 35 and older, and especially for

women with certain predisposing conditions. Women with histories of difficulties with blood clots are advised not to use contraceptive pills. However, it should also be noted that the incidence of deaths associated with pregnancy and labor is about four times greater than that associated with thromboembolic disease in users of the pill.

So serious are some of the pill's side effects that the Federal Food and Drug Administration now requires that each prescription for oral contraceptives contain a leaflet that describes the possible side effects as well as be labeled with a warning about smoking.

The warning about smoking is directed particularly at women 35 years of age and older. It is based on studies showing an increased risk of heart attack and other circulatory problems, such as stroke, for women who smoke while using oral contraceptives. Healthy women who do not smoke but who do use oral contraceptives double the risk of suffering a heart attack, compared with women who do not take the pill. However, users of the pill who also smoke, especially if they smoke 15 or more cigarettes a day, are three times more likely to die of a heart attack or of other circulatory disease than are users of the pill who do not smoke, and are ten times more likely to die of a heart attack or circulatory disease than are women who neither smoke nor use oral contraceptives. The risk of heart attack for women taking the pill increases with the amount of smoking and with age, and is higher in women with other conditions that may lead to heart attack, such as high blood pressure, high serum cholesterol levels, obesity, and diabetes.

In addition to the warning about smoking, the leaflet that now comes with each prescription of contraceptive pills points out several other possible dangers associated with the pill and makes the appropriate warnings:

[14]Two British studies reported higher death rates (Beral and Kay, 1977; Vessey, McPherson, and Johnson, 1977); but other researchers are critical of the applicability of these studies to the United States. For a complete discussion see Rinehart and Piotrow (1979) and Tietze (1979b).

Women who have had blood clotting disorders, cancer of the breast or sex organs, unexplained vaginal bleeding, a stroke, heart attack, or angina pectoris, or who suspect they may be pregnant should not take oral contraceptives.

Women with scanty or irregular periods (conditions that should be determined by a physician) are strongly advised to use another method of contraception, because if they use oral contraceptives they may have difficulty becoming pregnant or may fail to have menstrual periods after discontinuing the pill.

Possibly fatal (though rare) side effects include blood clots in the legs, lungs, brain, heart or other organs, cerebral hemorrhage, liver tumors that may rupture and cause severe bleeding, birth defects (if the pill is taken during pregnancy), high blood pressure, stroke, and gallbladder disease.

Estrogen, an ingredient in many oral contraceptives, causes cancer in certain animals and it may therefore also cause cancer in humans, though studies of women using the currently marketed pill do not confirm this. There is, however, evidence that estrogen use may increase the risk of cancer of the uterine lining in postmenopausal women [see Chapter 7].

Women who wish to become pregnant should stop using oral contraceptives and use a different method of birth control for a few months before attempting to become pregnant. This will help minimize the risk of birth defects associated with the use of sex hormones during pregnancy.

A woman should consult her physician before resuming use of the pill after childbirth, especially if she intends to breast-feed the baby, because hormones in the pill may be transferred to the infant through the mother's milk or may decrease the flow of milk.

The brochure that comes with oral contraceptives concludes: "Oral contraceptives are the most effective method, except sterilization, for preventing pregnancy. Other methods, when used conscientiously, are also very effective and have fewer risks. The serious risks of oral contraceptives are uncommon [rare] and 'the Pill' is a very convenient method of preventing pregnancy."

## Intrauterine Devices

Another method of contraception that has proved to be highly effective is the *intrauterine device*, or IUD. Such devices have had a long history. Ancient Greek writings mention IUDs, and for centuries Arab camel drivers have used IUDs to keep their camels from becoming pregnant on long journeys. The camel drivers inserted a round stone in the uterus of each camel before a journey, and this practice is still used in some parts of the world.

A variety of intrauterine devices for women was popular during the nineteenth century, both for contraception and in the treatment of such gynecological disorders as displacement of the uterus. These devices fell into disrepute in the early twentieth century, but in 1930 the German physician E. Gräfenberg developed a ring of coiled silver wire; he inserted duplicates in 600 women for the purpose of contraception. He reported a failure rate of only 1.6 percent. In the next few years, however, several gynecologists published articles condemning the use of intrauterine devices on the grounds that they might lead to serious infection of the pelvic organs. That some of the most vocal critics had no experience with this method did not lessen their influence, and intrauterine devices again fell into disrepute until 1959. In that year two promising reports were published, one by the Israeli W. Oppenheimer, who had worked with Gräfenberg, and one by the Japanese physician A. Ishihama. Oppenheimer reported having used the Gräfenberg device in 1500 women over a period of 30 years with no serious complications and a failure rate of only 2.4 percent. Ishihama reported on the use of a ring developed by another Japanese physician, T. Ota, which had been used in 20,000 women, with a failure rate of 2.3 percent and no serious complications. Ota was the first to use plastic instead of metal in an intrauterine device.

These reports triggered a new enthusi-

**Box 6.2** Early Contraception

People have been attempting to prevent unwanted pregnancies for millennia.[1] An Egyptian papyrus from 1850 B.C. prescribed various contraceptive methods. One consisted of crocodile dung mixed with a paste for use in the vagina. A mixture of honey and sodium carbonate was used to irrigate or plug the vagina.

Four centuries before Christ, Aristotle observed that "some prevent conception

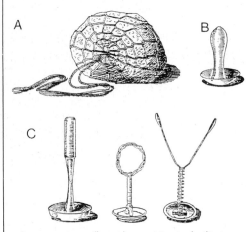

A sponge in a silk net has a string to facilitate withdrawal (above left). An intracervical pessary was used to plug the cervix (above right). Intrauterine pessaries extended into the uterus (below).

by anointing that part of the womb where the seeds fall with oil of cedar, or with ointment of lead, or with frankincense commingled with olive oil."

Many materials have been used to plug the cul de sac of the vagina and provide a barrier against sperm.[2] Some native Americans have used soft clay; and in other places leaves from various shrubs have been used. The Japanese have inserted paper; and the French, balls of silk. Various types of sponge and cotton balls have been used, sometimes in conjunction with medicated mineral oil or mild acid ointments to serve as spermicides. Many chemicals and drugs have been tried—with various levels of success—in different societies.

Intracervical pessaries made of gold, silver, or rubber have been used to plug the cervix. They were inserted by a doctor at the end of one menstrual period and removed at the onset of the next period. Intrauterine pessaries were longer and extended into the uterus to decrease the likelihood of their slipping out of place.

Needless to say, all of these methods were quite ineffective compared with modern contraceptives.

[1]Suitters (1967).
[2]Cooper (1928).

asm for intrauterine devices made of inert synthetic materials. By 1977 there were approximately 50 to 60 million IUDs in use throughout the world. China accounts for the largest number, with 40 million or more IUDs in use. Among married women in the United States of reproductive age about 6 percent use the IUD.

### Effects

How the IUD prevents pregnancies in humans is unknown. Although considerable research has been devoted to it, the mechanism is unclear. Studies have shown that IUDs are effective in every species of animal tested, but the mechanism varies from species to species. In sheep, for example, the IUD has been shown to inhibit the movement of sperm, thus preventing fertilization. In cows, on the other hand, it is the functioning of the corpus luteum that is impaired; and, though fertilization may occur, implantation in the uterus is inhibited. In the rhesus monkey a fertilized ovum can be found in a tube but not in the uterus of an animal fitted with an IUD, suggesting that the site of action is the uterus rather than the fallopian tubes.

In humans, studies have shown *no systemic effects* (for example, no alteration of pituitary or sex hormones secretion) from the IUD. This finding is in marked contrast with those related to the pill, which affects the pituitary glands, ovaries, breasts, uterus, liver, and other organs. Examination of the ovaries and fallopian tubes of women fitted with IUDs shows normal morphology and functioning. Tissues and secretions of the cervix and vagina are also normal.

At present the most widely accepted theory of the contraceptive action of the IUD is that it causes cellular and biochemical changes to occur in the uterus. It seems likely that the fertilized egg and sperm are destroyed or consumed by special cells of the body's defense mechanisms. A related theory is that the IUD interferes with implantation in some way that is still unknown.

### Types of IUDs

Intrauterine devices are made in various shapes and from various materials (Figure 6.1). A metal ring was the first widely used device, but it has the disadvantage that dilation of the cervix is required to insert or remove it. The same is true of the Ota ring, which is made of plastic and has been widely used in Japan and Taiwan. The Zipper ring is, however, made from coils of nylon thread and can be inserted through the cervix without prior mechanical dilation.

Most IUDs currently in use are made of plastic that is flexible and can be stretched into linear form to facilitate insertion. Once inside the uterus the device returns to its original shape. These devices all have nylon threads that hang down through the cervical opening into the vagina, enabling the wearer to check that the device is in place. The threads are small enough not to interfere with intercourse, however. These newer plastic

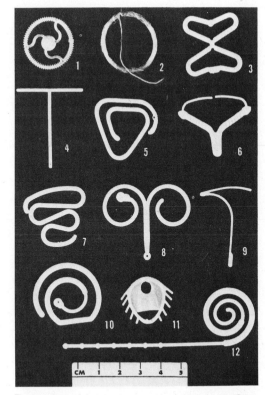

**Figure 6.1**  Various intrauterine devices: 1. Ota ring; 2. Zipper ring, 3. Birnberg Bow, 4. "T" device, 5. Ahmed, 6. K. S. Wing, 7. Lippes Loop, 8. Saf-T-Coil, 9. Copper "7," 10. New Margulies Spiral, 11. Dalkon-Shield, and 12. Gynecoil.  From Rudel, Kincl, and Henzl. *Birth Control: Contraception and Abortion*, figure 4.1 (New York: The Macmillan Company, 1973), p. 158. Copyright © 1973 by Macmillan Publishing Co., Inc.

devices also contain small amounts of barium in the polyethylene, which allows for visualization of the device on an X-ray picture.

Very popular among the new plastic devices is the Lippes loop. It comes in four sizes, ranging from 22.5 millimeters in diameter (for women who have not had children) to 30 mm. It is easily inserted by a physician, who stretches the loop into linear form and pushes it into the uterus through a plastic tube that is inserted in the cervical opening (*see* Figure 6.2).

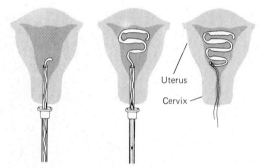

Uterus

Cervix

**Figure 6.2**   Insertion of the Lippes Loop.
Courtesy of Ortho Pharmaceutical Corporation.

Another relatively new device, the Copper 7, contains a small amount of copper that slowly dissolves in the uterus. (The amount of copper that dissolves daily is less than the amount recommended for a balanced diet and is therefore safe.) The Copper 7 is said to offer a higher rate of contraceptive protection than other IUDs, although how it does so is unknown. This device must be replaced every three years.

Finally, there is a recently developed IUD that contains progesterone, which is released at a slow, constant rate. The small amount of progesterone (.05 to .10 milligram per day) does not inhibit ovulation but alters the lining of the uterus in a way that prevents implantation. These progesterone-containing devices must be replaced every year.

**Effectiveness**

There is substantial agreement among various studies of the effectiveness of IUDs. Figures show a theoretical failure rate varying from 1 to 3 percent during the first year of use. In actuality the failure rate is approximately 5 percent. The failure rate tends to decline after the first year of use. One reason IUDs are usually more effective than some other contraceptive methods is that there is nothing either partner must remember to do to make it effective (although the woman

must check periodically to see that the device has not been expelled).

IUDs are also effective in preventing pregnancy in cases in which intercourse has taken place without any form of contraception being used.[15] Some doctors will insert IUDs under such circumstances within a few days after unprotected intercourse.

**Side Effects[16]**

The two most common side effects associated with IUDs are irregular bleeding and pelvic pain—seen in from 10 to 20 percent of the women using IUDs. However, as with most of the minor side effects associated with the pill, these problems tend to disappear after the first two or three months of use. Other side effects include bleeding, or "spotting," during the menstrual cycle and menstrual periods that may be heavier than usual after insertion of the IUD. Uterine cramps or general pelvic pain may also occur, and one study of 16,734 women reported that 833 had their IUDs removed for these reasons.[17] A lower incidence of cramping is reported with the progesterone containing IUD.

More serious and uncommon complications associated with the IUD are pelvic infection, perforation of the uterus, and problems during pregnancy. Normally these complications can be successfully treated, but in rare cases death can occur. The annual death rate associated with IUD usage is seven deaths per million users.

In the study mentioned above infection of the pelvic organs (uterus and tubes) was observed in 171 women, though hospital records showed that half of them had previous histories of such infections. As insertion of an IUD seems to exacerbate these conditions, women who have recently suffered pelvic infections are generally advised against using

[15]Lippes et al. (1978).
[16]For a more comprehensive review of this topic see Piotrow et al. (1979).
[17]Borell (1966), p. 53.

IUDs. Careful attention to antiseptic technique when inserting the IUD and careful screening for a history of prior infection can decrease the problem of infection. While it is more common for an infection to occur soon after insertion, it can occur two years or longer after insertion. The woman who has not had children appears to have a greater risk of infection than women who have had children. Perforation of the uterus is very rare (it occurs less than once in 1000 insertions). As with infection, the incidence of this complication varies with the skill and care of the physician inserting the device.

Women who accidentally become pregnant with the IUD in the uterus have an increased risk of several problems. Approximately 3 to 4 percent of these pregnancies will be ectopic. The rate for such pregnancies in nonusers is 0.4 percent. IUDs containing progesterone appear to be associated with a greater risk of ectopic pregnancy than other IUDs. If the pregnancy of the IUD user is a normal uterine pregnancy and the device is left in the uterus, approximately one-half of these pregnancies will spontaneously abort. This problem can be decreased by about 50 percent by removing the IUD. Also, if the device is left in place a rare complication, septic (infected) abortion, may occur. This problem can be avoided by removing the IUD when the pregnancy is discovered.

Spontaneous expulsion of IUDs can also be a problem, especially during the first year of use and especially with the smaller IUDs. In 1978 the FDA reported an overall expulsion rate of 4 to 18 percent during the first year after insertion.[18] Expulsion occurs most often in younger women and in women who have had no children and is more likely to occur during menstruation. When expelled into the vagina, the IUD is sometimes discarded with tampons or sanitary napkins

[18]*Second Report on Intrauterine Contraceptive Devices* (December 1978).

without the woman realizing it has been expelled. The various "tails" on IUDs enable women to check and verify that they are still in place.

There is no evidence so far that an IUD increases the risk of cancer of the cervix or uterus. Plastics of the type commonly used in IUDs have also been used extensively by surgeons for prosthetic devices in various parts of the body and have never been known to cause cancer. Some people have feared that a pregnancy that occurs with an IUD in the uterus might result in a deformed child. This is *not* true. The IUD does not cause an increased incidence of deformities.

### Subsequent Fertility

IUDs appear to have no effect on subsequent fertility and can be easily removed by a doctor when a woman wishes to become pregnant. Studies have shown that after removal of IUDs 60 percent of sexually active women become pregnant within three months and 90 percent within one year. These figures are similar to those for women who have never used any form of contraception. There is some concern that future studies *might* find a slightly decreased fertility in former IUD users due to the increased incidence of pelvic infection in women that use this device. In some cases pelvic infection can cause sterility.

## The Diaphragm

Until the advent of the pill and the intrauterine devices, the *diaphragm* and other mechanical devices designed to cover the cervix were widely used for contraception. Many women objected to the aesthetic or nuisance aspects involved in the repeated insertion and removal of these devices. Nevertheless, the absence of side effects related to the devices still outweigh the disadvantages as far as some women are concerned.

The diaphragm is the modern equivalent of an old idea. The history of contraception includes many examples of women in-

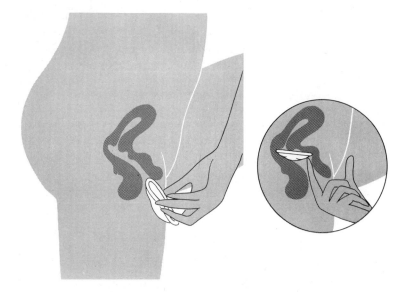

**Figure 6.3**  Insertion and placement of diaphragm. Courtesy of Ortho Pharmaceutical Corporation.

serting gums, leaves, fruits, seed pods, sponges, and similar items into their vaginas in attempts to block the sperm. The women of Sumatra, for example, molded opium into a cuplike shape and inserted it into the vagina. The women of Hungary used beeswax melted into round disks. The modern diaphragm was invented by a German physician in 1882.

### Effects

The principle of the diaphragm is straightforward. A thin rubber dome is attached to a flexible, rubber-covered metal ring; it is inserted in the vagina so as to cover the cervix and thus to prevent passage of sperm into the cervical canal (*see* Figure 6.3). The inner surface of the diaphragm (that is, the surface in contact with the cervix) is coated with a layer of contraceptive jelly before insertion.[19]

Diaphragms come in various sizes and must be individually fitted by a physician. They are usually about three inches in diameter. In addition, following the birth of a child or any other circumstance that may have altered the size and shape of her vaginal canal, a woman must be refitted.

The diaphragm and jelly must be inserted no more than six hours before intercourse in order for the jelly to be effective. Thus, unless a woman knows in advance that she is going to have intercourse on a particular occasion she may have to stop and insert the diaphragm in the midst of lovemaking or risk pregnancy. During menstruation it is unlikely that a woman will become pregnant. Nevertheless, some women use the diaphragm during menstruation to keep the lower portion of the vagina free of blood and to protect against the unlikely possibility of pregnancy.

When the diaphragm has been properly inserted, neither partner is aware of its presence during intercourse. After intercourse, the diaphragm should be left in place for six to eight hours and can be left in for as long as 24 hours. It should then be removed and washed. After drying it can be dusted with cornstarch if desired and stored in its plastic case. If it is reinserted, the contraceptive jelly must be applied.

[19]The effectiveness of the diaphragm is greatly decreased if the contraceptive jelly or cream is omitted.

### Effectiveness

In theory the diaphragm and spermicide together make a very effective method of contraception, with an ideal failure rate of 3 percent. The actual failure rate for the diaphragm is closer to 17 percent. There are several reasons for this discrepancy between the failure rates. First, the diaphragm is not always used when it should be or is not inserted correctly. Second, even if it is coated with jelly and correctly in place before intercourse, it may have slipped by the time of ejaculation. The diaphragm may become dislodged during intercourse and expose the cervix, especially if it fits loosely, or if the woman participates in vigorous intercourse with multiple withdrawals of the penis in a position other than flat on her back (particularly when she is on top of the man).[20]

The diaphragm is most effective when certain guidelines are followed:

The diaphragm and jelly can be inserted six hours before sexual intercourse; however, if a woman has intercourse more than two hours after insertion of the diaphragm, she should leave the device in place and insert an applicator full of spermicidal jelly or cream.

An additional applicator of cream or jelly should be inserted every time intercourse is repeated before the diaphragm is to be removed. The diaphragm is removed six to eight hours after the *last* act of intercourse.

The diaphragm should be refitted after pregnancy, a weight change of ten or more pounds, pelvic surgery, and one year after the start of intercourse on a regular basis.

## Spermicidal Substances

Contraceptive jellies, like those used with diaphragms, are among several *spermicidal,* or sperm-killing, *substances* available. Various foams, creams, jellies, and vaginal suppositories that kill sperm on contact are

[20]Calderone, ed. (1970), p. 234.

available in drugstores without a prescription and are simple to use. A plastic applicator is usually supplied for inserting the substance in the vagina. Vaginal foam, actually a cream packaged in an aerosol can, provides the best distribution of the spermicide in the vagina.

### Effectiveness

Of the currently available products, Delfen and Emko are among the best. Vaginal foams are most effective when the following directions are followed:

Insert a *full* applicator of foam as soon before intercourse as possible, but no longer than 15 minutes before intercourse. (Some clinics recommend using two full applicators.)

Insert the spermicide while lying down and do not get up prior to intercourse.

Do not douche for at least eight hours after intercourse.

Insert another applicator full with every act of intercourse.

Use the preparation with a condom for a very safe method of contraception.

The least effective spermicides are the foaming vaginal tablets and suppositories that get distributed more unevenly in the vagina and depend partly on mixing with natural lubricants and dispersion during the movements of intercourse. If ejaculation occurs before sufficient lubrication, mixing, and dispersion have occurred, the spermicide will be of little benefit. Another drawback is that the foam tablets often cause temporary irritation of the vagina. The theoretical failure rate for foams is 3 percent. The actual usage failure rate is closer to 22 percent. The creams and jellies are even less effective. The tablets and suppositories are the least effective with a failure rate closer to 30 percent.

A new vaginal suppository being sold in the United States is the *Encare Oval.* It is advertised as a very effective contraceptive, claiming a failure rate of 1 percent. It is highly questionable whether or not additional studies will confirm this low failure rate. Until further studies can be done on the effectiveness of this suppository, it should be considered to have a failure rate similar to the foam—about 20 percent.

## Douching

A time-honored but rather ineffective method of contraception is *douching,* that is, washing the sperm out of the vagina immediately after intercourse. This method is simple, requiring only a bidet or a douche bag and plain tap water. Various commercial products are available for douching; vinegar, lemon juice, soap, or salt may also be added to enhance the spermicidal properties of the solution. Actually these substances add little to the spermicidal properties of tap water and may irritate the vaginal tissues. The major disadvantage of douching as a contraceptive method is that within one or two minutes or *less* after ejaculation some sperm are already on their way up the cervical canal and beyond the reach of the douche. The woman must literally run from bed to bathroom if the douche is to be even mildly effective. The overall failure rate for the douche as a contraceptive method is about 40 percent.

## Condoms

Another contraceptive device that has been making a comeback in recent years is the *condom* (*see* Figure 6.4). Condoms, also known as "rubbers," "prophylactics," "French letters," and "skins," are thin, flexible sheaths worn over the erect penis to prevent sperm from entering the vagina. They are the only mechanical birth control device used by men. Condoms used to be kept hidden under the pharmacist's counter and sold rather quietly; now that they have become more popular, they are produced in bright colors and can be openly displayed in drugstores in most states. Approximately 750 million are sold in the United States each year.

Condoms are cylindrical sheaths with a ring of thick rubber at the open end. The thickness of the sheath is about 0.0025 inch (0.00635 centimeter). Each is packaged, rolled, and ready for use. Some condoms also come lubricated. The sizes of the various condoms are approximately the same.

The advantages of condoms include their availability and the protection they offer against venereal diseases. One minor disadvantage is that they reduce sensation somewhat and thus may interfere with the sexual pleasure of the male or female. (To some

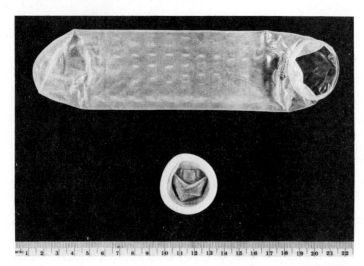

**Figure 6.4**  Condom, rolled and unrolled. Scale is in centimeters.

men this slight decrease in sensation might be desirable if it allows them to prolong sexual intercourse.) Putting on the condom interrupts sexual activity after erection but before the penis is placed in the vagina. This may be distracting, but many couples learn to integrate it smoothly into their sexual activity.

### Effectiveness

Condoms have been known to burst under the pressure of ejaculation, to leak, or to slip off during intercourse, but they are quite effective if used consistently. Failure rates range from a theoretical 3 percent (with very consistent usage) to an actual 10 percent. Using condoms with spermicidal cream or foam increases their effectiveness. Condoms are most effective when the following precautions are taken:

The condom should be put on before the penis comes in contact with the vulva because the secretions from the Cowper's glands may contain sperm that escapes before ejaculation.

When putting the rubber on, leave about one half inch (1.3 centimeters) at the end (if no reservoir is built in already) to allow space for the ejaculate so that it does not break the condom. The air should be squeezed out of the space at the end of the condom.

To avoid leakage the male should withdraw from the vagina soon after ejaculation—before detumescence—and should hold on to the rim of the rubber while withdrawing from the vagina.

Condoms (or the diaphragm) should not be used with vaseline or any other petroleum-based product that can destroy rubber. If additional lubrication is desired, a water soluble lubricant such as K-Y jelly may be used. The use of the contraceptive foam for lubrication would have the extra advantage of increasing the safety of the condom.

Store condoms away from heat. Unopened condoms remain good for about two years if kept away from heat.

## Withdrawal

*Withdrawal* of the penis from the vagina just before ejaculation (*coitus interruptus*) is probably the oldest known method of birth control and is still commonly used throughout the world.[21] The decline of the birthrate in western Europe from the late nineteenth century onward is believed to have been due to the popularity of this method.

The major problem with coitus interruptus is that it requires a great deal of motivation and willpower just at the moment when a man is most likely to throw caution to the winds. Nevertheless, this method costs nothing, requires no devices, and has no physiological side effects—although some people find it psychologically unacceptable.

When withdrawal is the only contraceptive measure taken, the theoretical failure rate is 9 percent. Actual failure rates range from 20 to 25 percent. This is partly because the male does not always withdraw quickly enough or far enough from the vulva and partly because small amounts of semen may escape before ejaculation.

## The Rhythm Method

The *rhythm method* involves abstaining from intercourse during the presumed fertile period of the menstrual cycle. (*see* Chapter 4). This method is unreliable, particularly for women whose menstrual cycles are irregular. If a woman has kept track of her periods for 10 or 12 months she can calculate her fertile period from Table 6.1. The "safe period" for intercourse includes only those days not included in the "fertile period."

---

[21]As we mentioned at the beginning of this chapter there is reference to this method in *Genesis* 38:8–9: "Then Judah said to Onan, 'Go in to your brother's wife, and perform the duty of a brother-in-law to her, and raise up offspring for your brother.' But Onan knew that the offspring would not be his; so when he went in to his brother's wife he spilled the semen on the ground, lest he should give offspring to his brother."

**Table 6.1** The Fertile Period

| Shortest Cycle (Days) | Day Fertile Period Begins |
|---|---|
| 22 | 4 |
| 23 | 5 |
| 24 | 6 |
| 25 | 7 |
| 26 | 8 |
| 27 | 9 |
| 28 | 10 |
| 29 | 11 |
| 30 | 12 |
| 31 | 13 |
| 32 | 14 |
| 33 | 15 |
| 34 | 16 |

| Longest Cycle (Days) | Day Fertile Period Ends |
|---|---|
| 22 | 12 |
| 23 | 13 |
| 24 | 14 |
| 25 | 15 |
| 26 | 16 |
| 27 | 17 |
| 28 | 18 |
| 29 | 19 |
| 30 | 20 |
| 31 | 21 |
| 32 | 22 |
| 33 | 23 |
| 34 | 24 |

The safe and unsafe days are calculated by subtracting 18 from the woman's shortest cycle and 10 from her longest cycle. Thus, for the woman with a regular 28-day cycle the fertile period extends from day 10 through day 18. For women with cycles ranging from 24 to 32 days, the periods of abstinence must extend from the sixth through the twenty-second day, an unacceptable span for many couples. Nevertheless, to have intercourse during this period is akin to playing Russian roulette, for there is no way of knowing in advance whether or not "the chamber is loaded."

There are several ways of making the rhythm method more effective. One is based on changes in a woman's body temperature, which goes up slightly at the time of ovulation. In order to pinpoint the time of ovulation and fertility by this method a woman must take her temperature immediately upon awakening every morning before arising, moving about, eating, drinking, or smoking. An increase of 0.4°F (0.2°C) above the average temperature of the preceding five days indicates ovulation if the increase is sustained for three days. Minor illnesses like colds and sore throats, however, can throw off the temperature curve. Also, some women do not have their temperature rise even though they have ovulated.[22]

A BBT (basal body temperature) chart does indicate when ovulation has occurred, but it is not helpful before ovulation. So if the BBT chart (*see* Figure 6.5) is to be used to determine the "safe period," a woman should abstain from intercourse from the end of her menstrual period until three days after the time of ovulation.

A device called the ovutimer has been developed by researchers at the Massachusetts Institute of Technology. It can be used to make the rhythm method more reliable and can also be used by couples who want to know when the fertile days are so they can try to conceive a child. The ovutimer is a 7-inch (17.8-centimeter) long plastic device that, when inserted into the vagina, determines the time of ovulation by measuring the stickiness of cervical mucus, which becomes thin and watery at the time of ovulation.

The ovutimer and the BBT method promise to make the rhythm method more reliable, but even when the time of ovulation is known, the success of the rhythm method depends on a couple's ability and motivation to follow directions exactly. Large-scale studies show actual failure rates ranging from 15

[22]Moghissi (1976).

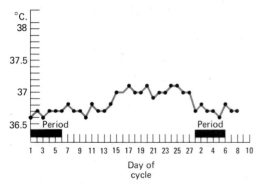

°C.
38
37.5
37
36.5
Period                Period

1  3  5  7  9  11 13 15 17 19 21 23 25 27 2  4  6  8 10

Day of
cycle

**Figure 6.5**   BBT chart showing ovulation about the thirteenth day (36°C = 98°F). From *International Planned Parenthood Federation Medical Handbook*, p. 62. Reprinted by permission of the International Planned Parenthood Federation, 18–20 Lower Regent Street, London SW1, England.

to 35 percent when the rhythm method is the only contraceptive method used.

## Cervical Mucus

To use the *cervical mucus method* a woman must watch the changes in the cervical mucus that occur during the monthly cycle in order to determine when intercourse is safe or unsafe.[23] She can study the mucosa by checking her toilet paper and underpants or by putting a finger into her vagina. Some women insert a speculum (a device for separating the vaginal walls) into the vagina in order to judge the state of the mucus. The changes in the cervical mucus should be written down every day on a calendar. For a few days after menstruation many women have a few "dry" days during which no noticeable discharge is present and there is a sensation of dryness in the vagina. Some women have no dry days after menstruation. The "dry" days are considered relatively safe for intercourse. Within a few days after menstruation is over a mucus discharge is produced which is white or cloudy and tacky. Gradually the mucus

[23]This method has been developed by Evelyn and John Billings who have described it thoroughly in their book, *Atlas of the Ovulation Method*.

changes to be clear and stretchy—rather like the consistency of egg white. This is called the peak symptom and usually lasts for one or two days. Generally, a woman ovulates around 24 hours after the last peak symptom day. After the last peak symptom day the mucus changes to cloudy and tacky. Intercourse must not occur beginning with the first day in which mucus is present until four days after the last peak symptom day. By that time the egg will no longer be capable of fertilization and a woman can engage in intercourse without becoming pregnant. The actual failure rate for this method appears to be approximately 15 to 25 percent. At present the research on the effectiveness of this method is limited.

## Sterilization

Voluntary surgical sterilization has become increasingly popular in recent years among both men and women in the United States and in countries like India, where sterilization (especially for males) has been encouraged by the government. In fact, the acceptance rate of sterilization by people in the United States in the last half of their reproductive years (older than age 29) has increased significantly in recent years. It is estimated that more than one million surgical sterilizations are now being done in the United States each year.

### Male Sterilization

The operation that sterilizes the male is the *vasectomy,* a simple procedure that can be done in a doctor's office in about 15 minutes. A small amount of local anesthetic is injected into each side of the scrotum, and a small incision is made on each side in order to reach the vas deferens (*see* Chapter 2). Each vas is then tied in two places, and the segment between is removed in order to prevent the two cut ends from growing together again. After this operation sperm will no longer be able to travel through the vas from the testes. The sperm, which continue to be produced but are now trapped in the testes, are simply reabsorbed by the body as they degenerate.

No change in sexual functioning occurs as a result of vasectomy. The sex glands continue to function normally, secreting male sex hormones into the blood. Ejaculation still occurs because the seminal fluid contributed by the testes through the vas only accounts for about 10 percent of the total volume. The only difference is that the semen will be free of sperm. Sperm may still be present two or three months after a vasectomy because they are stored in the reproductive system beyond the site of the vasectomy, but these sperm can be flushed out with water or with a sperm-immobilizing agent during the vasectomy. Once these remaining sperm are gone, vasectomy is virtually 100 percent effective as a birth control measure.

Vasectomy does not interfere physiologically with a man's sexuality. His sexual response and orgasm after vasectomy are the same as before. Some men feel a new sense of freedom after being sterilized, but others experience negative psychological effects— possibly because vasectomy is usually considered a permanent form of sterilization. However, improved surgical techniques are changing this, and vasectomy may eventually become a reversible or temporary form of sterilization. The reversal procedure is not simple, but the cut ends of the vas can be reunited surgically. Using pregnancy of partner as a measure of success, some surgeons report rates of reversal of vasectomy as low as 18 percent, others as high as almost 80 percent.[24] Another method of making vasectomy reversible consists of inserting a small valve in the vas. The valve allows the flow of sperm to be turned on or off when desired. These valves are still in the experimental stage, and the truly reversible vasectomy is still in the future, but there are several reasons

why some men request that their vasectomy be undone: remarriage after divorce or death of wife, death of one or more children, improved economic condition, and removal of negative psychological effects of vasectomy.

## Female Sterilization

The most common surgical procedure for sterilizing women is often called "tying the tubes" or *tubal ligation.* Tying or cutting the fallopian tubes prevents eggs from reaching the uterus and sperm from reaching the eggs in the fallopian tubes. The eggs that continue to be ovulated are simply reabsorbed by the body as they degenerate. Female sterilization used to be a major surgical procedure involving hospitalization, general anesthesia, and all the associated costs and risks. The search for simple, effective, and inexpensive sterilization procedures, however, has led to the development of more than a hundred techniques for cutting, closing, or tying the tubes.

There are still medical situations in which a major abdominal operation may be necessary for female sterilization, but the current trend is in the direction of an outpatient procedure performed under local anesthetic. It is possible to approach the fallopian tubes through the vagina rather than through the abdominal wall and to perform the sterilization with a *culdoscope.* This instrument is basically a metal tube with a self-contained optical system that allows the physician to see inside the abdominal cavity. Sterilization using the procedure called *culdoscopy* involves puncturing the closed end of the vagina and, after locating the tubes with the culdoscope, tying and cutting them.

In addition to the traditional methods of tying the tubes, a variety of clips, bands, and rings have been developed for blocking the tubes. Chemicals that solidify in the tubes, caps that cover the ends of the tubes, and lasers that heat and destroy a portion of the tubes are among the sterilization methods currently under investigation. Various plastic and ceramic plugs have also been designed

[24]Silber and Cohen (1978).

**Table 6.2**  Summary of Contraceptive Methods

| Method | Ideal Failure Rate[1] | Actual Failure Rate[1] | Advantages | Disadvantages |
|---|---|---|---|---|
| Birth control pills | 0.34% | 4–10% | Easy and aesthetic to use | Continual cost, side effects; requires daily attention |
| IUD | 1–3% | 5% | Requires little attention; no expense after initial insertion | Side effects, particularly increased bleeding; possible expulsion |
| Diaphragm with cream or jelly | 3% | 17% | No side effects: minor continual cost of jelly and small initial cost of diaphragm | Repeated insertion and removal; possible aesthetic objections |
| Vaginal foam | 3% | 22% | Easy to use: no prescription | Continual expense |
| Douche | ? | 40% | Inexpensive | Inconvenient; possibly irritating |
| Condom | 3% | 10% | Easy to use: helps to prevent venereal disease | Continual expense; interruption of sexual activity and possible impairment of gratification |
| Withdrawal | 9% | 20–25% | No cost or preparation | Possible frustration |
| Rhythm | 13% | 21% | No cost, acceptable to Roman Catholic Church | Requires significant motivation, cooperation, and intelligence; useless with irregular cycles and during postpartum period |
| Cervical mucus | ? | 15–25%[2] | No cost | Requires careful observation, cooperation |
| Vasectomy | 0.15% | 0.15% | Permanent relief from contraceptive concerns | Possible surgical/medical/psychological complications |
| Tubal ligation | 0.04% | 0.04% | Permanent relief from contraceptive concerns | Possible surgical/medical/psychological complications |

[1]Hatcher et al. (1978).
[2]Klaus, et al. (1977); Wade et al. (1979); World Health Organization (1978).

for blocking the tubes, and one still-experimental plug appears to be both effective and removable—allowing for reversibility of sterilization.

Most methods currently being used to sterilize women are almost 100 percent effective and, like vasectomy, have no physiological effect on sexual functioning. Female sterilization, like male sterilization, is considered a permanent method of birth control. But, recent advances have resulted in an increased success at reconnecting the tubes. One study reports a pregnancy rate of 80 percent after reversal.[25] Generally, the rates of pregnancy after reversal are 25 to 50 percent depending on which method of sterilization was used and the skill of the doctor.

In addition to sterilization by tying the tubes, many women have been sterilized by the surgical procedure known as *hysterectomy* (surgical removal of the uterus). It is estimated that one-third of all women in the United States have had a hysterectomy by age 65, but charges have been made that the operation is overused. The operation is performed in one of two ways: through an incision in the abdominal wall or through the vagina. The ovaries are left in place (unless there is some medical reason for their removal), so the secretion of female sex hormones remains normal. Hysterectomy is usually not performed solely as a means of sterilization, however, but is done because of some medical problem, such as tumor removal.

See Table 6.2 for a comparison of the contraceptive methods discussed in this chapter.

[25]Ibid.

## Abortion

*Abortion* has been used as a form of birth control for thousands of years in numerous cultures, whether or not the procedure was considered legal. In the United States in the mid-1800s, for instance, abortion was relatively common: It is estimated that there was one abortion for every five or six live births. Abortionists advertised in newspapers and frequently sold drugs that were supposed to induce abortion. Laws were gradually enacted against abortion in the last century, but these laws were not always strictly enforced. Illegal abortions continued to be performed. Kinsey found that about 23 percent of the white women he sampled had an illegal abortion by the time they finished their reproductive years. Illegal abortions are much more dangerous to the woman than legal abortions are. In 1973 abortion was legalized in the United States, and by 1977 legal abortions were estimated at 1.3 million per year. Meanwhile, illegal abortions declined from an estimated 530,000 in 1970 to 10,000 per year.

Even though legal, abortion has not become a primary means of birth control. As a backup procedure when contraception fails, however, abortion is becoming increasingly popular, and although it remains a highly controversial issue on ethical and moral grounds (*see* Chapter 15), it is widely practiced in the Soviet Union, parts of eastern and central Europe, and Japan. It is less common in some of the Catholic countries of Europe and South America.

The method used for abortion in the United States is usually determined by the length of the pregnancy. During the first trimester, abortion is performed by mechanically removing the contents of the uterus through the cervix. Evacuation is sometimes used as late as the twentieth week, but during the second trimester, abortion is usually performed by stimulating the uterus to expel its contents—in effect, inducing a miscarriage.

Although abortion does present risks,

the overall death rate due to abortion is relatively low. The death rate for legal abortions performed during the first trimester is 1.7 deaths per 100,000 abortions. Almost 90 percent of abortions in the United States are performed during the first trimester. In the second trimester the figure increases to 12.2 per 100,000. This death rate is slightly higher than the death rate due to childbirth complications: 11.2 per 100,000.

### Vacuum Aspiration

We have already mentioned (*see* Chapter 4) the use of suction techniques for "menstrual extraction." *Vacuum aspiration* has become very popular for first-trimester abortions for a number of reasons. It can be performed on an outpatient basis quickly and at a relatively low cost. Prior to eight weeks of pregnancy, minimal or no anesthesia may be required, and it is often unnecessary to dilate the cervix mechanically.

Following preparation similar to that for a pelvic examination a suction curette is passed through the cervical opening into the uterus. Suction curettes come in various shapes and sizes, but most are essentially a plastic tube with a hole at each end and a hole on the side. One end is inserted into the uterus (*see* Figure 6.6) and the other end is attached to a suction pump. The hole on the side can be covered and uncovered as required to increase or decrease the vacuum pressure within the uterus. The suction tip is rotated gently within the uterus until, on examination, all tissue related to conception has been eliminated. The curette is then removed and, in some clinics, a uterus-contracting drug (for example, oxytocin; *see* Chapter 5) may be administered to minimize bleeding and ensure complete evacuation of the uterus. The entire procedure (excluding examination and preparation time) can usually be performed in less than two minutes.

Complications of vacuum aspiration are relatively uncommon, but may include per-

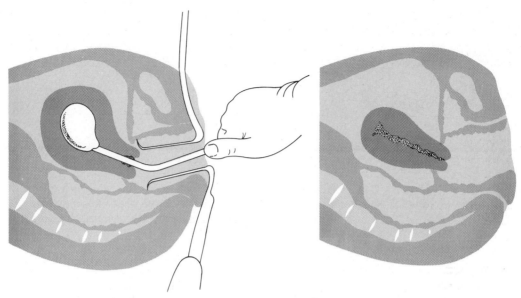

**Figure 6.6**    (*Left*) initial insertion of angled suction curette into the cervical canal; (*right*) uterus contracted after completion of evacuation of the uterus.

foration of the uterus, hemorrhage, uterine infection, cervical lacerations, and drug reactions, if anesthetics or other drugs are used. The specific incidence of these complications varies with the skill and experience of the physician, as well as the state of health of the woman undergoing the abortion.

### Dilation and Curettage (D and C)[26]

Another method of abortion sometimes used during the first trimester is dilation of the cervix and curettage (scraping) of the uterus. The first step, *cervical dilation,* can be accomplished by passing a series of progressively larger metallic dilators (curved rods) through the cervical opening, but in recent years a less painfull (if slower) method has become popular—*laminaria sticks.* These sticks, which are made from compressed seaweed, are inserted into the cervix. As they absorb cervical secretions they expand to five times their dry

size in about 24 hours. When the cervix is enlarged sufficiently, a *curette* (a bluntly serrated metal instrument) is inserted and used to scrape the tissues off the inner walls of the uterus. In more advanced pregnancies dilation may have to be more extensive in order to allow passage of a forceps that can be used mechanically to grasp the fetal tissues for removal. The possible complications of an abortion by the D-and-C method are the same as those of vacuum aspiration.

### Dilation and Evacuation (D and E)

After the twelfth week of pregnancy, abortion becomes an increasingly serious procedure, and the rate of complications increases. One abortion technique that is often used between the thirteenth and twentieth weeks of pregnancy is *dilation and evacuation.* D and E is similar to D and C and the suction method, although the fetus is larger at this stage and not as easily removed as during the first trimester. Once dilation is achieved, suction, forceps, and curettage are used. D and

---

[26]D and C is also employed for the diagnosis and treatment of a number of pelvic disorders.

---

## Box 6.3 New Female Contraceptives

A variety of promising new contraceptive methods are being developed for use by women.

Plastic vaginal rings that are impregnated with progestins have been proved to be 98 percent effective in preventing ovulation.[1,2] The rings are slightly smaller than a regular diaphragm and are more easily inserted. They are placed in the vagina on the fifth day of the menstrual period and are left in place for the next 21 days. Menstruation begins a few days after the ring is removed. Each ring can be used up to six months. The progestins in the rings are released slowly and absorbed through the vaginal wall into the woman's bloodstream. The hormones prevent ovulation, change the endometrium, and thicken the cervical mucus. There are few side effects, and the ring does not interfere with intercourse.

Instead of taking a pill every day, it is possible that a safe, long-acting injection will be developed for contraceptive purposes. One such preparation is already being used in some parts of the world but has not been approved for use in the United States.[2] Each dose of the preparation (brand name, Depo Provera) contains 150 milligrams of progesterone, enough to prevent ovulation for 90 days.

Side effects include irregular menstrual bleeding and an unpredictable period of infertility following cessation of the shots. The FDA's primary reason for disapproval of this drug for long-term contraception is based on studies showing that it is related to an increased incidence of premalignant breast tumors in dogs. The drug's effect on humans is still being investigated.

HCG—human chorionic gonadotropin—is the hormone produced by the zygote that travels to the woman's body and signals: "You're pregnant." HCG causes the corpus luteum to continue producing estrogen and progesterone; and these two hormones prevent menstruation while preparing the endometrium for the early phases of pregnancy. Antibodies to HCG are being developed which allow the female body to be immunized to its own HCG.[1] Women with antibodies to HCG no longer respond to the HCG mechanism that signals the body to skip menstruation and to begin the hormonal developments of pregnancy. When antifertility vaccines become available, immunized women will menstruate each month, whether their ova are fertilized or not.[2]

[1]Shapiro (1977)
[2]Hatcher et al. (1978)

---

E is considered to be a simple, effective, and relatively inexpensive procedure, and recent studies suggest that, up to the twentieth week, D and E is as safe as or safer than some of the other procedures that are commonly used during the second trimester.

### Saline Abortions

During the second trimester, abortion is commonly induced by the injection of a concentrated salt solution into the uterus. It should be noted that *saline abortion* is more complicated for the woman than is abortion in the first trimester. The fourth month of pregnancy is a particularly difficult time for an abortion. The pregnancy is too far along to allow for a safe, simple aspiration; but the uterus is not yet large enough to allow the physician to easily locate the proper place in the abdominal wall in which to insert a needle for saline injection. Because of this problem many physicians are reluctant to perform saline abortions earlier than the sixteenth week of pregnancy.

## Box 6.3    Continued

Another promising new method for the future is a zinc-medicated and acidically buffered collagen sponge.[3] Collagen is a natural protein prepared from bovine skin and Achilles tendons. Collagen has a very high fluid binding capacity and the collagen sponge can easily absorb several ejaculates. The device is shaped somewhat like a diaphragm and inserted into the upper vagina where it covers and blocks the cervical opening. The acidity of the vagina and the sponge destroys the sperm which become trapped in the sponge. In addition, the sponge also has some spermicidal effects because of the zinc additives. The sponge can be removed after 8 hours from intercourse and washed. It can then be reinserted and left in place for 3 to 5 days since it does not irritate the vaginal mucosa. The collagen sponge has a beneficial side effect: the zinc appears to prevent genital herpes from replicating. For best results, it is used in conjunction with spermicides.

[3]Chvapil (1976, pers comm)

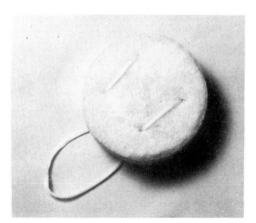

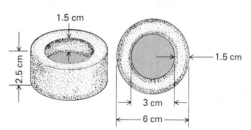

The collagen sponge and applicator. Courtesy of Dr. M. Chvapil, Section of Surgical Biology, College of Medicine, University of Arizona.

The mechanism of action of the saline abortion is not known. The method is quite straightforward. After examination (to verify exact location of the uterus) and sterile preparation of the abdominal skin, a needle is inserted through the abdominal and uterine walls using local anesthesia. About 200 cubic centimeters (less than seven ounces) of amniotic fluid is removed (compare the procedure for amniocentesis described earlier). An injection of 200 cubic centimeters of a 20 percent sodium chloride (common salt) solution is then introduced into the uterus. Some clinics send the woman home to await the onset of uterine contractions, while others keep all saline abortion patients hospitalized for observation until the abortion is complete. Contractions usually begin within 12 to 24 hours, and about 80 percent of women deliver the fetus and placenta within 48 hours of the injection. Others take longer and may require another injection of saline. Some physicians administer oxytocin to stimulate more vigorous labor contractions.

---

**Box 6.4**  New Male Contraceptives

At present males have only three methods they can use to prevent unwanted pregnancies: withdrawal, the condom, and vasectomy. What changes are likely to occur in the next decade or two? Expert opinions vary considerably. Some researchers believe that effective methods will be available within a few years. Others predict that nothing will be available until after the year 2000.[1]

There are promising leads and many researchers are working on male contraceptive methods; but the problems of controlling millions of sperm without damaging other cells, causing genetic mutations, or reducing testosterone output are considerable. The current guidelines on clinical experimentation and drug testing in the United States will probably introduce long time lags between the development and wide usage of any new male method.

The Chinese have reported very rapid progress in developing and testing a male method that they claim is 99.8 percent

effective and has no serious side effects. At first it was noted that men had decreased fertility in certain areas of China in which unrefined cottonseed oil was part of the daily diet. Gossypol in the cottonseed oil appears to interfere with sperm production in the seminiferous tubules, resulting in nonviable sperm. For the past several years, thousands of Chinese men have been using gossypol for birth control, with apparent success. In addition, the men have been reported to regain fertility after they stop taking gossypol. This contraceptive method would not be so rapidly adopted in the United States because gossypol is known to have several toxic effects, hence extensive testing would be needed.

Numerous researchers have been studying other drugs. Almost every chemical that reduces sperm count has also had toxic effects on the body. However, some drugs—such as danazol—have had

[1]Djerassi (1979).

---

Uncommon but severe complications are known to occur with saline abortion. Most serious is *hypernatremia* ("salt poisoning"). Early symptoms of hypernatremia are abdominal pain, nausea, vomiting, and headache. More serious results, if correct treatment is not instituted promptly, can include high-blood pressure, brain damage, and death.

Other complications of saline abortion include intrauterine infection and hemorrhage. Delayed hemorrhage (days or weeks after the abortion) occurs most commonly in association with the complication of retained placenta—that is, in instances where the fetus is successfully aborted but part or all of the placenta remains behind. In such cases curettage and blood transfusion may sometimes be required.

**Prostaglandin Abortions**

More than 45 years ago it was reported that injections of human semen into the uterus cause vigorous contractions of uterine muscles. By 1935 it was shown that the substance responsible for this action is produced by the prostate gland and seminal vesicles, and it was thus named *prostaglandin*. Later it was discovered that there are more than a dozen chemically related prostaglandins, and some of them can now be synthesized in the laboratory.

Because prostaglandins cause uterine contractions, they can be used to induce abortion. They are usually injected directly into the uterus but can also be injected into the bloodstream or into a muscle. Laminaria sticks are sometimes used with prostaglandins

**Box 6.4**   Continued

promising results in reducing male fertility.[2] Danazol is a synthetic hormone with a structure similar to the progestins. It prevents the release of FSH and LH from the pituitary and it tends to have masculinizing effects on the body, rather than feminizing effects. In preliminary studies, men have been given danazol in conjunction with testosterone for 6 months at a time. The drug

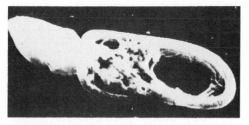

A sperm coated with T-mycoplasmas (× 12,900). From Fowlkes, DM, Dooher, GB, O'Leary, WM: Evidence by scanning electron microscopy for an association between spermatozoa and T-mycoplasmas in men of infertile marriage. *Fertility and Sterility* 26:1203, 1975. [Fig. 5, p. 1207]. Reproduced with the permission of the publisher, The American Fertility Society.

caused sperm counts to drop to 0.5–5 percent of normal levels in the majority of men tested. The men did not lose sexual function, and they regained normal fertility within 5 months after the end of treatment.

A microorganism called T-mycoplasmas has been linked to infertility in some men who are naturally infertile. In contrast to normal sperm, the sperm from men with T-mycoplasmas are coated with spherical T-mycoplasmas microorganisms and have coiled tails.[3] The sperm from the men with T-mycoplasmas cannot move as fast as sperm from men without the microorganism. Men who have T-mycoplasmas can become fertile again after receiving antibiotics that destroy the microorganisms. Because T-mycoplasmas and the active chemical from it cause infertility, they may offer possible new approaches to male contraceptive methods.

[2]Shapiro (1977:289).
[3]Fowlkes et al. (1975a, b).

because by dilating the cervix they reduce the number of uterine contractions necessary to expel the fetus. This can hasten the abortion of the fetus and the placenta. Some success has also been reported in bringing about abortion after insertion of a prostaglandin suppository into the vagina. This procedure has been used to induce first- and second-trimester abortions, but suction is still the primary procedure for first-trimester abortions in the United States.

Complications with prostaglandin abortion include nausea, vomiting, and headache, with at least 50 percent of women experiencing one or more of these side effects. These effects are temporary, rarely serious, and can be easily treated. Complications such as hem-

orrhage, infection, and uterine rupture (possible with all types of second-trimester abortion) are infrequent with prostaglandin abortion. One major drawback is that the proportion of live births is higher with prostaglandin than with saline abortion, especially after the twentieth week (45 live births out of 607 abortions, in one recent study). For this reason some physicians do not like to perform abortions after the twentieth week. When they do, they sometimes use saline, and some have reported success in using a combination of prostaglandin and urea. Urea can also be used alone to induce abortion. It has the same effects as saline but is slightly less effective than either saline or prostaglandin.

# Chapter 6

# Sexual Disorders

Since everyone experiences physical ill health and mental turmoil at one time or another, it is realistic to view a certain amount of illness as a natural part of life. In addition to the ordinary wear and tear on our bodies and minds, we are all subject to certain common ailments, as well as to some less common and more serious disorders. Some knowledge of these matters is helpful in allaying worry and discomfort as well as in alerting us to the presence of danger signals.

A vast number of disorders may affect the sex organs and sexual functioning. The sex organs may also suffer secondary effects from general systemic diseases. Several medical specialties are involved in the treatment of these various disorders. *Gynecologists* deal with disorders of the female sexual or reproductive system. Most male disorders are treated by *urologists*.

We shall deal primarily with three types of sexual disorder in this chapter: first, venereal diseases and other sexually transmitted diseases; second, common bothersome conditions; and third, certain serious conditions the early recognition of which may save lives.

## Sexual Hygiene

We wish to mention, in connection with disorders of the reproductive system, some general principles for the cleansing and care of the genitals and related tissues. Specific preventives of disease, if these are available, are mentioned in conjunction with specific diseases later in this chapter.

It is fair to say that any person who practices regular and thorough cleansing of the genitals and related areas is less likely to be afflicted with some of the infectious diseases we will describe. As we will discuss in Chapter 10, one's attractiveness as a sexual partner may be enhanced by specific attention to matters of sexual hygiene. It is not by any means necessary to equate hygiene with the elimination or covering up of natural odors. A large-scale advertising campaign in recent years has attempted to convince people that vaginal deodorant sprays and similar products are a necessity for sexual hygiene. Actually, these products should be avoided as some of them have been found to contain chemical irritants, and there may also be allergic reactions associated with their use in some people. There is a difference, however, from both hygienic and aesthetic viewpoints, between the causes and effects of fresh and stale odors. The fresh and often attractive odor of someone who bathes regularly is related to a combination of the secretions of various glands located on and around the genitalia and, perhaps, a slight residue of unscented or mildly scented soap. The stale and often offensive odor of someone who bathes infrequently, or does not wash the genital region carefully, is related to the action of skin bacteria and other microorganisms on accumulated body secretions and, on occasion, remnants of fecal material as well. A lengthy delay in changing tampons or externally worn absorbent pads during menstruation can produce similar results, but this is a less common occurrence.

Most of the skin is smooth, relatively hairless, and relatively devoid of glands other than sweat glands. Both the male and female genitalia, on the other hand, are wrinkled (the labia in the female, the scrotum and foreskin in the male), surrounded by hair, and rich in various glands (*see* Chapter 2). These three factors contribute to the likelihood of unpleasant odors in this area. (The next most likely source of strong odor is under the arms, where similar conditions exist, except that there are not as many crevices that can be overlooked.) The primary means of preventing these adverse health and aesthetic consequences is simple: regular, careful washing of the genital region with soap and water. "Regular" means several times a week or more, at least as frequently as one has sexual intercourse (and, from the partner's viewpoint, preferably *before* intercourse.) "Careful" means somewhat methodical coverage of the genitals with a washcloth; a simple shower is less than satisfactory for cleansing this area.

Although the vagina is a self-cleansing organ, some women do find it desirable to cleanse the vagina by douching. This is not usually necessary for hygienic purposes, but the insertion of creams, foams, gels, lubricants, and so forth results in residues that may necessitate douching for some of the women who use them. Others prefer to douche at least once a month at the end of their menstrual periods. If a woman does choose to douche, she normally should not do so more than twice a week. Aside from possibly being irritating, excessive douching can upset the vaginal environment and may lead to infection. Plain tap water or a mildly acidic solution (for example, two tablespoons of vinegar in

a quart of water) is preferable to most commercial douching preparations. Regular, alkaline (soda) douches should be avoided because they interfere with the normal chemical balance and flora of the vagina.

With these general principles in mind, let us move to a discussion of the causes, treatment, and prevention of various sexual disorders.

## Venereal Diseases

The term "venereal" comes from the Latin *venereus*, "pertaining to Venus," the goddess of love. Venereal diseases are spread through intimate sexual contact. Our discussion will focus on *gonorrhea*, the most common form of "VD" in the United States, and *syphilis*, the most serious. Less common and less serious venereal diseases are mentioned later in this chapter.

Although both diseases are readily curable with penicillin or other antibiotics, the incidence of gonorrhea is rising, though at a slower rate than between 1965 and 1975 (*see* Figure 7.1). The incidence of syphilis has been relatively stable since the early 1960s (*see* Figure 7.2). The preponderance of cases occurs in the 15–29 age group (*see* Figure 7.3); it is currently estimated that 50 percent of American young people contract syphilis

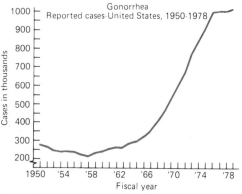

**Figure 7.1**   Gonorrhea, reported cases, United States, 1950–1978.

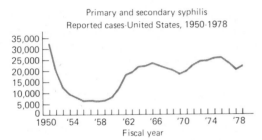

Primary and secondary syphilis
Reported cases-United States, 1950-1978

**Figure 7.2**  Primary and Secondary Syphilis, Reported cases, United States, 1950–1978.

or gonorrhea by age 25. It is not an overstatement to describe the present situation in the United States as an epidemic. In recent years gonorrhea has been the most prevalent disease of those reported to the U.S. Public Health Service, and syphilis ranks third. Gonorrhea is more common today than measles, mumps, or tuberculosis. Of all the contagious diseases, VD is second in prevalence only to the common cold. Although public health codes require physicians to report all cases of venereal disease to public health authorities, in actual practice probably less than one-third of the cases are reported.[1] For instance, there

[1]Kolata (1976).

were 1,013,436 *reported* cases of gonorrhea in 1978 (*see* Figure 7.1), but experts estimate that there were probably more than 3 million cases that year.

The purpose in reporting cases of venereal disease is to enable public health workers to locate sexual "contacts" of contagious individuals and to treat them. This method has been successful in virtually eliminating epidemics of other infectious diseases. But many people are unwilling to disclose names and addresses of sexual contacts for a variety of reasons, ranging from embarrassment to fear of more serious consequences, including self-incrimination and incrimination of past or present lovers. In many states intercourse with someone other than a spouse is a criminal offense. Homosexual acts are also crimes in most states (*see* Chapter 14). Although such information cannot legally be used as a basis for criminal prosecution, it is not difficult to understand the reluctance of someone who, by revealing the identity of a sexual partner, is revealing the identity of a partner in crime.

## Gonorrhea

Gonorrhea is an infection caused by the bacterium *Neisseria gonorrhoeae*, and can affect

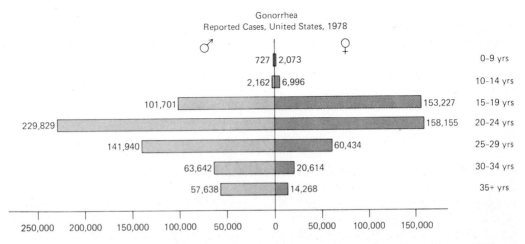

**Figure 7.3**  Age distribution for reported cases of gonorrhea, United States, 1978.

## Box 7.1    Gonorrhea Evolving

Ever since 1958, new strains of gonorrhea have been appearing that are resistant to penicillin. Originally, 200,000 units of penicillin were needed to cure gonorrhea. Today, nearly 5 million units are needed, an increase of 25 times! As doctors utilize higher doses, the least resistant forms of gonorrhea are destroyed, while the most resistant forms survive. Over the years, this process of selection has caused the evolution of increasingly resistant forms of gonorrhea.

A new form of gonorrhea—called "supergon" or "superclap"—has now appeared that is different from the penicillin-resistant forms.[1] The new strain of gonorrhea has the ability to produce the enzyme penicillinase, which destroys penicillin and renders the drug totally ineffective in controlling the disease.

Penicillinase-producing gonorrhea appears to have originated in the Philippines, where prostitutes regularly take low doses of penicillin to keep from contracting venereal diseases. Under these conditions, any new form of gonorrhea that could destroy penicillin would have a selective advantage in the "survival of the fittest." It appears that as many as 20 to 40 percent of the Philippine prostitutes now carry the new penicillinase-producing gonorrhea. Military personnel and the merchant marine have facilitated the spread of the new disease to other countries. By early 1976, penicillinase-

producing gonorrhea had been reported in 18 countries.

The first case of penicillinase-producing gonorrhea was reported in the United States in 1975. By June 1979, 685 cases had been reported in 36 states, the District of Columbia, and Guam; and the disease is continuing to spread. Penicillinase-producing gonorrhea can be cured by the antibiotic spectinomycin. However, there are two problems. First, spectinomycin costs 7 or 8 times as much as penicillin. Although people in affluent countries can afford the increased cost, it is likely that the drug will not be widely used in less-developed countries. Second, when spectinomycin is used with greater frequency in treating gonorrhea, it is likely that new forms of gonorrhea will evolve that are resistant to this drug. Drug researchers are working on new drugs to replace obsolete drugs; but scientists are concerned about staying ahead of the new forms of the various diseases that are developing resistance to current drugs.

At present people with gonorrhea are treated with penicillin. Only if a patient has proved resistant to penicillin is spectinomycin used as a treatment. If spectinomycin were used more extensively, it would only hasten the evolution of new forms of gonorrhea with resistance to that drug.

[1]Culliton (1976).

a variety of mucous-membrane tissues. This microorganism does not survive without the living conditions (temperature, moisture, and so on) provided by the human body. It is transmitted from human being to human being during contact with infected mucous membranes of the genitalia, throat, or rectum. A new strain of gonorrhea that has evolved recently is potentially more dangerous than the original forms of gonorrhea (see Box 7.1).

Ancient Chinese and Egyptian manuscripts refer to a contagious urethral discharge that was probably gonorrhea. The ancient Jews and Greeks thought that the discharge represented an involuntary loss of semen. The Greek physician Galen (A.D. 130–201) is credited with having coined the term "gonorrhea'' from the Greek words for ''seed'' and ''to flow.'' For centuries gonorrhea and syphilis were believed to be the same disease, but by the nineteenth century a series of experiments had demonstrated that they were two separate diseases. In 1879, A. L. S. Neisser identified the bacterium that causes gonorrhea and now bears his name (*see* above).

### Symptoms in Males

In males the primary symptom of gonorrhea (known also as the ''clap'' or ''strain'') is a purulent, yellowish urethral discharge. The usual site of infection is the urethra, and the condition is called *gonorrheal urethritis*. Most infected males have symptoms, but a few males are asymptomatic. A discharge from the tip of the penis usually appears within three to ten days after contraction of the disease, and is often accompanied by burning during urination and a sensation of itching within the urethra. The inflammation may subside within two or three weeks without treatment, or it may persist in chronic form. The infection may spread up the genitourinary tract to involve the prostate gland, seminal vesicles, bladder, and kidneys. In 1 percent of cases the disease spreads to the joints of the knees, ankles, wrists, or elbows, causing gonorrheal arthritis, a very painful condition.

More than 90 percent of cases clear up immediately with prompt penicillin treatment. For persons allergic to penicillin, tetracyline or erythromycin may be used. The discharge often disappears within 12 hours after treatment, although a thin flow persists for a few days in 10–15 percent of patients.

Gonorrheal urethritis can usually be prevented by one of two methods (*see* Box 7.2): (1) use of a condom and thorough washing of the sex organs and genital area with bactericidal soap or solution after sexual exposure, or (2) a single dose of penicillin or other appropriate antibiotic within a few hours after exposure.

### Symptoms in Females

In females the symptoms of gonorrhea may be mild or absent in the early stages. In fact, 80 percent of women with gonorrhea (and 10 percent of men) are essentially without symptoms, a major factor in the unwitting spread of the disease. The primary site of infection is usually the cervix, which becomes inflamed (*cervicitis*). The only early symptom may be a yellowish vaginal discharge which may be unnoticed by the female. Not all such discharges are gonorrheal, however. Microscopic examination and bacterial culture of the discharge are required for definitive diagnosis.[2] Treatment with antibiotics is usually effective if the disease is recognized and treated promptly.

If left untreated, however, the infection may spread upward through the uterus to involve the fallopian tubes and other pelvic organs. Often this spread occurs during menstruation, when the uterine cavity is more susceptible to gonorrheal invasion. Acute symptoms—severe pelvic pain, abdominal distension and tenderness, vomiting, and fever—may then appear during or just after menstruation. Again, treatment with antibiotics usually brings about a complete cure, but if the disease is not treated or is inadequately treated, a chronic inflammation of the uterine tubes (*chronic salpingitis*) ensues. This condition is accompanied by formation of scar tissue and obstruction of the tubes and constitutes a common cause of infertility in females, particularly those who frequently contract gonorrhea.

## Nongenital Gonorrhea

*Pharyngeal gonorrhea* is an infection of the throat that is transmitted most commonly during fellatio (oral stimulation of the penis). Cunnilingus (oral stimulation of the vulva)

[2]Unfortunately, there is no routine blood test that will detect gonorrhea. This is a major reason why identification of asymptomatic gonorrhea has been much less successful than identification of asymptomatic syphilis.

**Box 7.2** Controlling Gonorrhea

In the United States the number of cases of gonorrhea reported in the 1970s increased at an alarming rate. It nearly doubled between 1969 and 1976. In contrast, the number of cases reported in Sweden has been declining! It dropped by 50 percent between 1969 and 1976.[1]

# Kondom

The Swedish condom symbol.

The major reason for the decline in gonorrhea in Sweden appears to be increased use of the condom. Because gonorrhea is passed between the male and female through the male's urethral meatus, covering the penis with a condom blocks the route of transmission. Condoms also provide partial protection against the transmission of syphilis, herpes, and other sexually transmitted diseases by covering the penis. The Swedish government has encouraged use of the condom through public education and by making the devices more readily available. The Swedish condom symbol (illustrated) has been displayed widely in order to heighten people's awareness of the condom's value. As condom use increased, gonorrhea rates decreased. In Denmark, Norway, and Finland—neighboring countries in which there was no program of public education about the effectiveness of condoms in controlling gonorrhea—there was no decrease in gonorrhea rates.

Condoms are rarely mentioned on television in the United States. Newspapers, magazines and even radio have been more candid about venereal diseases than TV.[2] Because many people do not want to read or hear about venereal disease, the media have been reluctant to deal with these topics.

[1]Kolata (1976).
[2]Blakeslee (1976).

does not usually cause pharyngeal gonorrhea. (Kissing does not provide sufficient contact to transmit gonorrhea.) The primary symptom is a sore throat, but there may also be fever and enlarged lymph nodes in the neck. In some cases there may be no symptoms.[3]

*Rectal gonorrhea* is an infection of the rectum usually transmitted during anal intercourse. In women with gonorrhea of the cervix the infection is sometimes spread to the rectum by exposing the rectum to the infected vaginal discharge.[4] Both rectal gonorrhea and pharyngeal gonorrhea are common in male homosexuals. The symptoms are itching associated with a rectal discharge.[5] Many cases are mild or asymptomatic, however. Treat-

[3]Fiumara (1971), p. 204, and Wiesner (1975).

[4]Rein (1975).
[5]Schroeter (1972), p. 31.

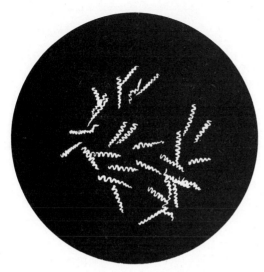

**Figure 7.4** Typical organisms of *Treponema pallidum* from tissue fluid in a dark field. The length of each is about ten microns. From Jawetz et al., *Review of Medical Microbiology,* 9th ed. (Los Altos, Calif.: Lange Medical Publications, 1970), p. 220. Reprinted by permission.

ment of rectal or pharyngeal gonorrhea is the same as for gonorrheal urethritis.

Until recently a common cause of blindness in children was *ophthalmia neonatorum,* a gonorrheal infection of the eyes acquired during passage through infected birth canals. Instilling penicillin ointment or silver nitrate drops into the eyes of all newborn babies is now compulsory and has helped to eradicate this disease.

## Syphilis

It was not until 1905 that the microorganism that causes syphilis was identified. A German investigator, Fritz Richard Schaudinn, identified and named *Spirochaeta pallidum,* describing it as a "slender, very pale, corkscrew-like object" (*see* Figure 7.4). (The name *Treponema pallidum* is now the technically correct one, although "spirochete" is more commonly used.)

### Symptoms

In its late stages syphilis can involve virtually any organ or tissue of the body, producing myriad symptoms similar to those of other diseases, which led the famous physician Sir William Osler to call it the "Great Imitator." The first stage or *primary stage,* is marked by a skin lesion known as a *chancre* at the site where the spirochete entered the body. The chancre (pronounced "*shank-er*")

**Figure 7.5** A chancre of the penis. Note the raised, hard appearance of the ulcer. From Dodson and Hill, *Synopsis of Genitourinary Disease,* 7th ed. (St. Louis: C. V. Mosby Co., 1962), p. 201. Reprinted by permission.

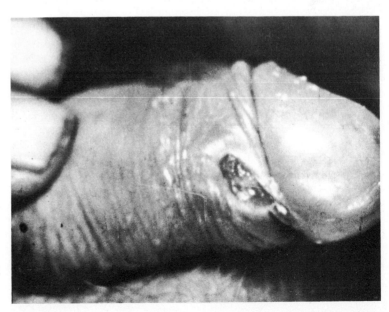

## Box 7.3  The Spread of Syphilis

There are two theories about the origins of syphilis. Some believe that syphilis was present in the Old World since ancient times but became more virulent, perhaps because of a mutation, in the late fifteenth century.

It is more commonly believed that syphilis was brought to Europe by Columbus and his crew after their first voyage to the West Indies. It is true that, within a few years after Columbus' return in 1493 from his first voyage to the New World, epidemics of syphilis spread across Europe with devastating effects. History suggests that the Spaniards introduced the disease to the Italians while fighting beside the troops of Alfonso II of Naples. Then in 1495 an army of mercenaries fighting for Charles VIII of France conquered Naples. As they returned home through France, Germany, Switzerland, Austria, and England, they took the disease along with an excess of celebration. By 1496 syphilis was rampant in Paris, leading to the passage of strict laws banishing from the city anyone suffering from it. In 1497 all syphilitics in Edinburgh were banished to an island near Leith. In 1498 Vasco de Gama and his Portuguese crew carried the disease to India, and from there it spread to China; the first epidemic in that country was reported in 1505. Outbreaks of syphilis in Japan later followed the visits of European vessels.

The New World origin of syphilis is also suggested by the discovery of definite syphilitic lesions in the bones of Indians from the pre-Columbian period in the Americas. There are no comparable findings in the bones of ancient Egyptians, nor are there any clear descriptions of the disease in the medical literature of the Old World before Columbus. Physicians in the early sixteenth century did not have a name for the disease, but the Spaniards called it the "disease of Española" (present-day Haiti). The Italians called it "the Spanish disease"; and the French called it "the Neopolitan disease." As it spread to many countries it acquired the name *morbus Gallicus,* the "French sickness," a name that persisted for about a century. The term "syphilis" was introduced in 1530 by the Italian physician Girolamo Fracastoro, who wrote a poem in Latin about a shepherd boy named Syphilus (from the Greek *siphlos,* meaning "crippled" or "maimed"), who caught the disease as a punishment from the gods (for having insulted Apollo). As this name was ethnically neutral, it gradually became accepted as the proper term for the dread disease.

Various historical figures have been afflicted by syphilis, as is indicated by records of their physical appearances and symptoms. Columbus himself died in 1506 with symptoms typical of advanced syphilis, involving the heart, extremities, and brain. It is generally accepted that the first four children of Catherine of Aragon, first wife of Henry VIII of England, all died of congenital syphilis, leaving only one survivor, the future "Bloody Mary." (Mary died at age 42—of complications of congenital syphilis, it appears.) Henry's disappointment over not having a male heir undoubtedly played a role in his insistence on legalizing his second and subsequent marriages, which led to the break between England and Rome.

is a hard, round ulcer with raised edges, and is usually painless. In the male it commonly appears somewhere on the penis, on the scrotum, or in the pubic area (*see* Figure 7.5). In the female it usually appears on the exter-nal genitals (*see* Figure 7.6), but it may appear in the vagina or on the cervix and thus escape detection. It may also appear on the mouth, in the rectum, on the nipple, or elsewhere on the skin.

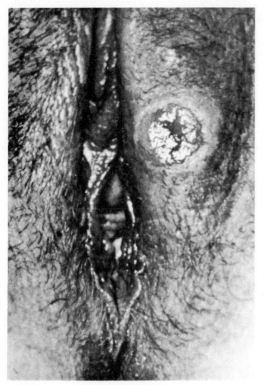

**Figure 7.6** A large chancre on the labia majora. Primary syphilis in the female is not usually this obvious. From Weiss and Joseph, *Syphilis* (Baltimore: Williams & Wilkins, 1951), p. 73. Copyright © 1951 by The Williams & Wilkins Co.; reprinted by permission.

As syphilitic infections in men may begin at sites other than the penis, condoms do not necessarily provide protection against them. And, as it is usually not apparent that someone is a carrier of syphilis, the only sure way to avoid the disease is to know before sexual contact whether or not the anticipated partner has had a negative blood test for syphilis—an unlikely question to ask as a prelude to making love. The point is, however, that casual sexual contacts are more likely to expose one to syphilis (or other venereal disease), for statistically the greater the number of sexual partners, the greater the probability of encountering someone who is a carrier.

The chancre appears two to four weeks after contraction of the disease and, if not treated, usually disappears in several weeks, leading to the illusion that the individual has recovered from whatever he or she thought was the trouble. Actually this chancre is usually only the first stage in the development of a chronic illness that may ultimately be fatal. Treatment with penicillin or other antibiotics when the chancre occurs cures most cases, and relapses after proper treatment are rare.

### Secondary and Tertiary Syphilis

When syphilis is untreated, the *secondary stage* becomes manifest anywhere from several weeks to several months after the healing of the chancre. There is usually a generalized skin rash, which is transient and may or may not be accompanied by such vague symptoms as headache, fever, indigestion, sore throat, and muscle or joint pain. Many people do not associate these symptoms with the primary chancre.

Syphilis (and gonorrhea) are transmitted *only* by intimate contact with another human being. The most infectious times are when a chancre is present or during the second stage of the disease when skin lesions are present (particularly lesions in moist areas of the body, such as the mouth). Thus, syphilis can be caught through kissing if the chancre is on the lips or if the lesions are on the mouth. Explanations involving contact with contaminated toilet seats, wet towels, chairs, drinking glasses, swimming pools, or domestic animals may save face but are pure myth and/or self-deception.

After the secondary stage all symptoms disappear, and the so-called *latent period* begins. During this period, which may last from 1 to 40 or more years, the spirochetes burrow into various tissues, particularly blood vessels, the central nervous system (brain and spinal cord), and bones.

About 50 percent of untreated cases reach the final, or "tertiary," stage of syphilis,

## Box 7.4 Three Uncommon Venereal Diseases[1]

There are three venereal diseases that are relatively uncommon in the United States. There is a higher incidence of these diseases in tropical areas; and they are more prevalent in the southern states than in the North.

Chancroid ("soft chancre") is caused by a bacillus known as *Hemophilus ducreyi* (the bacillus of Ducrey). The primary lesion of this venereal disease is a chancre which resembles the syphilitic chancre in appearance, but, in contrast to the syphilitic lesion, is quite painful. Diagnosis is based on microscopic examination or culture of the bacillus. Treatment with sulfa drugs is quite effective. The disease has become quite rare in Western countries, although it is still common in the East and in tropical regions. In 1978, there were 521 cases reported in the United States.

*Lymphogranuloma venereum* ("tropical bubo," LGV) is caused by a microorganism that is neither a bacterium nor a virus but has some of the properties of each. Although the site of entry is usually the penis, vulva, or cervix, the first obvious manifestation of the disease is usually enlarged, tender lymph glands in the groin accompanied by fever, chills, and headache. Treatment consists of sulfa or broad-spectrum antibiotics such as chlortetracycline. LGV is most common in the tropics and subtropics. In the United States it is seen most frequently in the South. In 1978, 284 cases were reported in the United States.

*Granuloma inguinale* ("chronic venereal sore") is caused by an infectious agent known as *Donovanian granulomatis*, or the "Donovan body," after its discoverer. Like LGV, it is most common in the tropics and subtropics. It does occur in the southern U.S. states; seventy-two cases were reported in the United States in 1978. The disease is characterized by ulcerated, painless, progressively spreading skin lesions. It is not highly contagious. The most common sites of infection are the skin and mucous membranes of the genitalia, but the disease may also involve the rectum, buttocks, or mouth. The most effective antibiotics for treatment are tetracycline and streptomycin.

[1]Brown *et al.* (1970); King and Nicol (1975).

---

in which heart failure, ruptured major blood vessels, loss of muscular control and sense of balance, blindness, deafness, and severe mental disturbances can occur. Ultimately the disease can be fatal, but treatment with penicillin even at late stages may be beneficial, depending on the extent to which vital organs have already been damaged.

Syphilis can be transmitted to the fetus through the placenta; hence the mandatory blood tests to identify untreated cases of the disease before marriage and before the birth of a child. Treatment with penicillin during the first half of pregnancy can prevent congenital syphilis in the child. Nine out of ten pregnant women who have untreated syphilis either miscarry, bear stillborn children, or give birth to living children with congenital syphilis.

Children with congenital syphilis are prone to impaired vision and hearing, as well as to certain deformities of the bones and teeth. Treatment with penicillin can alleviate many of the manifestations of congenital syphilis if it is initiated early in infancy.

## Other Sexually Transmitted Diseases

In recent years doctors and public health workers have become increasingly aware of the sexual transmission of other diseases besides the traditional venereal diseases. In the past these diseases were less common or the degree to which they were sexually transmitted was not realized. While many of the following diseases have become quite common, doctors and laboratories are not required to report these diseases to Public Health Service officials. Thus, there are no national figures on the incidence of these diseases although estimates are available based on limited samples.

## Trichomoniasis

A common vaginal infection is *trichomoniasis,* which is caused by a protozoan called *Trichomonas vaginalis.* It is characterized by an odorous, foamy, yellowish or greenish discharge that irritates the vulva, producing itching or burning sensations. Trichomoniasis is usually a sexually transmitted disease.[6] A man may harbor this organism in the urethra or prostate gland without symptoms, or he may have a slight urethral discharge. Since both sexual partners may have the infection, it is customary to treat both partners simultaneously with a drug called metronidazole (*Flagyl*) to prevent reinfection.[7] Trichomonads are sometimes spread nonsexually. If an uninfected person's genitals come in contact with something such as a washcloth or wet bathing suit that has been contaminated with the protozoa through contact with the genitals of an infected person, the uninfected person may become infected.

## Nongonococcal Urethritis

A form of urethritis that is perhaps more common than that caused by gonorrhea is *nongonococcal urethritis* (NGU). This disease is most often seen among white and affluent patients and is probably the most common form of urethritis seen in student health centers and in the offices of private physicians.

NGU is transmitted sexually (the organism involved is frequently, but not always, *Chlamydia trachomatis*), and although it causes urethritis in males (with a urethral discharge as in gonorrhea), it may be asymptomatic in females.

Because NGU was only recently identified as a sexually transmitted disease and because it is sometimes asymptomatic, many cases have gone undiagnosed and mistreated. The result is that this disease is becoming more and more widespread. NGU can be effectively treated with tetracycline and usually clears up within five days of treatment.

## Nonspecific Vaginitis

*Nonspecific vaginitis* occurs when women have vaginitis (irritation of the vagina) but the organism responsible for the infection cannot be isolated or identified easily. As a result, many cases have been undiagnosed and mistreated. Some forms of nonspecific vaginitis are sexually transmitted. The frequency of nonspecific vaginitis is difficult to determine, but health officials believe that it is becoming more widespread.

Nonspecific vaginitis can be caused by the same organisms that cause nongonococcal urethritis in the male. As already mentioned, *Chlamydia trachomatis* can affect both sexes. *Hemophilus vaginalis* is probably a common cause of nonspecific vaginitis.[8] Because it can often be recovered from the sexual partners of infected women, it is possible that it is sexually transmitted. Ampicillin is the treatment of choice for women infected with *Hemophilus.*[9]

## Herpes

One viral infection of the genitals is *Herpes Simplex Virus Type 2* (HSV-2), which is now recognized as being a very common sexually

---

[6]Rein and Chapel (1975).
[7]There have been highly publicized reports that large doses of this drug may cause tumors in rats. There is no evidence of such effects in humans.

[8]McCormack (1974).
[9]Rein and Chapel (1975).

transmitted disease in the United States. The Public Health Service estimates that there are 300,000 new cases of genital herpes a year. HSV-2 should not be confused with HSV-1, which causes "cold sores," eye infections, and skin conditions above the waist. Some cases of genital herpes (perhaps 10 percent) are caused by HSV-1, however. Likewise, some herpes infections above the waist are caused by HSV-2.

The HSV-2 infection results in the appearance of small vesicles (fluid-filled pockets or blisters) surrounded by inflamed tissue, which usually appear three to seven days after exposure to the virus. The most common sites of herpes infections in females include the surface of the cervix, the clitoral prepuce, and the major and minor lips; in males, the foreskin and glans. When the blisters are internal, such as on the cervix, the condition may go undetected because the cervix is insensitive to pain. In some instances, males may also be asymptomatic. Herpes causes burning and itching sensations in most cases. When the blisters break open, they may become infected with bacteria from the skin.

The usual mode of transmission of HSV-2 is intimate sexual contact; therefore sex should be avoided when herpes blisters or ulcers are present. Even though the blisters usually clear up after a few weeks, the virus may remain in the body for years, and the infection can recur cyclically. Although some people never have another active outbreak, others have recurring outbreaks for years.

There are special problems associated with herpes. A pregnant woman who has an active outbreak at the time of birth runs a risk (one in four) that her newborn child will be seriously damaged or die of the virus. To avoid this problem many doctors will perform cesarean sections in this situation. In rare cases the fetus has become infected while still in the uterus.[10] There is also the possibility

that HSV-2 can cause cancer. Approximately 6 percent of women who have HSV-2 develop cervical cancer, so it is recommended that women who have had herpes get Pap smear tests every six months to check for cancer.

One reason that the disease has spread so rapidly is that people often do not recognize it or know that they have it. Also, in some instances the sores associated with recurring outbreaks may not be noticed during a casual inspection.

People with active herpes are advised to take hot sitz baths and to keep sores clean and dry. Other treatments for herpes have not been found to be effective. But in 1979 a new method for treating herpes was developed that appears promising.[11] The drug used in this treatment, 2-deoxy-D-glucose, is an antiviral agent that interferes with the multiplication of herpes virus. In a study on 36 women with active genital herpes, 89 percent of the cases were cured within four days (*see* Figure 7.7). Of course, further studies are needed to guarantee its efficacy.

## Genital Warts

*Genital warts* (also known as *condylomata acuminata*) are most commonly transmitted by sexual contact, although this is not their only mode of transmission. The symptoms include obvious bumps and possible itching or irritation. They usually appear from one to three months after contact. Genital warts are similar to common plantar warts and both are caused by a virus. They appear most often in women on the vulva, vagina, and, less frequently, the cervix; in males, on the surface of the glans and below the rim of the corona. They have also been known to develop in and around the anus, particularly in male homosexuals. Traditionally, physicians have treated genital, or "venereal," warts in several ways: surgically, with electrodes, with cryosurgery (freezing), or with the direct ap-

[10]Kaufman and Rawls (1974).

[11]Blough and Giuntoli (1979).

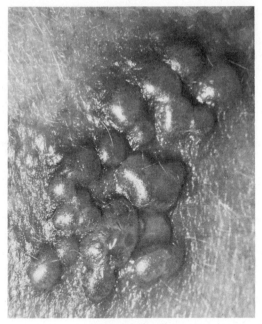 

**Figure 7.7**  Typical herpes virus lesions before treatment (*left*) and four days afterward, after cure (*right*). From Blough and Giuntoli, Successful treatment of human genital herpes infections with 2-deoxy-D-glucose, *Journal of the American Medical Association*, 241 (© 1979), 2798–2801. Courtesy of Dr. H. A. Blough and the American Medical Association.

plication of various drugs. A vaccine made from the patient's own wart tissue has also been shown to be effective in fighting the virus and in getting rid of the warts.

## Pubic Lice

*Pediculosis pubis* (pubic lice, "crabs") is an infestation of the pubic hair and is usually transmitted sexually. However, it is occasionally acquired from contact with infested bedding, towels, or toilet seats. The primary symptom of pubic lice is intense itching, which results from bites (the lice feed on blood like mosquitoes). Adult lice are visible to the naked eye, but just barely. They are bluish-gray and about the size of a pinhead. Cream, lotion, or shampoo preparations of gamma benzene hexachloride (*Kwell*) are very effective in eliminating both adult lice and their eggs ("nits").

## Scabies

Scabies is a contagious skin infection that is caused by an itch mite (*Sarcoptes scabiei*). The mites provoke intense itching and can be spread by close personal contact, including sexual contact and contact with infested objects such as clothing or bedding. Scabies is commonly found in the genital areas, buttocks, and elsewhere on the body. The female itch mite burrows into the skin and lays eggs along the burrow. The larvae hatch within a few days. The mites are more active and cause the most itching at night. Once the person begins to scratch the infected areas, the scratching usually opens the skin to bacterial infection which causes secondary inflammation. This parasitic infestation can be eliminated by washing with *Kwell*, the same preparation used in treating lice.

## Other Common Problems

### Leukorrhea

Some amount of vaginal mucus is normal. Women have varying amounts of discharge during their monthly cycle and during sexual excitement. When the discharge becomes excessive or abnormal it is called *leukorrhea*. Almost *every* woman experiences leukorrhea at some time in her life. It is not a discrete disease entity but rather a condition that can be caused by infections, chemicals, and physical changes. We have already mentioned several infections such as trichomoniasis that can cause leukorrhea. Irritating chemicals in commerical douche preparations may also cause vaginal discharges. In fact, frequent douching of any sort is likely to increase the production of vaginal mucus. Leukorrhea may result from irritation by foreign bodies (such as a contraceptive device). Leukorrhea may also be related to alterations in hormone balance (during pregnancy or menopause).

### Candidiasis

A common vaginal infection that causes irritation and a discharge is *candidiasis* (also known as moniliasis, monilia, or yeast). It is caused by a yeastlike organism called *Candida albicans.* The thick white discharge causes itching and discomfort, which may be severe. This organism is normally present in the vagina, but it produces an infection only when it multiplies excessively. Candidiasis is most commonly seen in women using oral contraceptives, in diabetic women, and during the course of pregnancy or prolonged antibiotic therapy for some other condition. Candidiasis can be very annoying, especially when it involves itching skin on the thighs. Usually this condition responds to treatment with nystatin (*Mycostatin*) suppositories. Though much less common in males, this fungus infection does occur, particularly under the foreskin of an uncircumcised male. Often these males have no symptoms, but there may be a white discharge and irritation. An infected male is also treated with nystatin, but in cream form.

### Prostatitis

*Prostatitis*, or inflammation of the prostate gland, is a relatively common problem in men who have passed the age of 40.[12] Chronic prostatitis is much more common than acute prostatitis. The acute form of the disease may result from complications of either nonspecific urethritis, gonorrhea, or other infections.[13] Although most cases of the acute form can be resolved with antibiotic treatment, some cases become subacute or chronic.

Chronic prostatitis can be caused by infections, prolonged congestion of the posterior urethra due to excessive intake of alcohol, prolonged sexual excitement without ejaculation, or long periods of abstinence after times of frequent ejaculation.[14] Among the symptoms of chronic prostatitis are lower back pain, perineal pain, incomplete erection, urinary problems, and premature, painful, or bloody ejaculation. In the past, physicians often recommended that men with prostatitis should abstain from sexual activity and ejaculation. Today, it is more often recommended that men should continue their normal rate of sexual activity, although they should avoid prolonged foreplay without ejaculation in order to minimize congestion of the gland. Massage of the prostate has been used to help drain the ducts of the gland, but sexual release may be more effective. Antibiotics may be of use in some cases; but complete cure is often not possible. Warm sitz baths may reduce the symptoms.

### Cystitis

*Cystitis* is an infection or inflammation of the bladder. It is not a sexually transmitted disease, but its occurrence in women is sometimes associated with sexual activity. (It is so

[12]Barnes et al. (1967).
[13]Sturdy and Lyth (1974).
[14]Davis and Mininberg (1976).

common among newlywed women, for instance, that the term "honeymoon cystitis" has been used to describe it.)

The bacteria that invade the bladder through the urethra are not usually caught from the sexual partner, but are normally present on the genital skin of the infected person. Women are more prone to cystitis because their urethras are significantly shorter than those of males. The primary symptom of cystitis is frequent and painful ("burning") urination. It may subside spontaneously in a few days, but it is advisable to receive proper antibiotic treatment because untreated infections may spread from the bladder to the kidneys, causing a much more serious condition called *pyelonephritis*.

## Cancer of the Sex Organs

Common cancers in both sexes involve organs of the reproductive system.

### Of the Breast

Cancer of the breast[15] is the most common form of cancer in women. It is extremely rare in women under age 25, but increases steadily in each decade thereafter. For women in the 40–44 age group, cancer of the breast is the most common cause of death. Ultimately, almost 1 in 13 women will develop breast cancer. Males too can develop breast cancer, although this condition is rare, accounting for less than 1 percent of all breast cancers. The cause of this disease is unknown, but it is clear that blows to the breast do not cause breast cancer. Many breast cancers respond to sex hormones. The spread of these cancers may be accelerated by increased hormone secretion during pregnancy. Some investigators feel that the long-term ingestion of hormones in birth control pills may stimulate the growth of breast cancer. Thus far the available re-

search does not support this hypothesis. Studies dealing with the effects of the pill on breast cancer have not found an increase in breast cancer among women who use oral contraceptives. More research is being conducted in this area.

Removal of the ovaries is often beneficial in the treatment of breast cancer, as is treatment with the male sex hormone testosterone. The primary treatment is surgical removal of the breast (*mastectomy*) and related tissue.

When breast cancer is detected and breast removal recommended, one of the most difficult problems the woman faces is the disfigurement caused by the mastectomy. But this situation is changing. Breast reconstruction techniques have become so sophisticated in recent years that women no longer need fear loss of self-image and confidence as a result of the operation. Plastic surgeons not only can reconstruct the breast contours (using silicone gel implants), they can also reconstruct the nipple and areola from the intact nipple.

Cancer of the breast can be fatal, but with early diagnosis and treatment the prognosis is much more favorable. About 65 percent of patients with cancer of the breast are still alive five years after the initial diagnosis. Early diagnosis and treatment are often missed because cancer of the breast begins with a painless lump in the breast, which may go unnoticed for a long time, in part because approximately three-fourths of women do not regularly perform a breast self-examination. This is unfortunate since the best way to reduce the mortality rate for breast cancer is to detect the disease early before it spreads to other parts of the body. Over 90 percent of all breast cancers are discovered by the women themselves. Thus, every woman beginning in her teens should do a breast self-exam once a month about one week after the end of her period or on one set day a month after menopause (*see* Box 7.5).

---

[15]Although not technically sex organs, the breasts are part of the reproductive system and are highly sensitive to sex hormones, as was noted in Chapter 4.

**Box 7.5** How To Do a Breast Self-Exam[1]

1. **In the Shower.** Examine your breasts during your bath or shower, since hands glide more easily over wet skin. Hold your fingers flat and move them gently over every part of each breast. Use the right hand to examine the left breast and the left hand for the right breast. Check for any lump, hard knot, or thickening.

2. **Before a mirror.** Inspect your breasts with arms at your sides. Next, raise your arms high overhead. Look for any changes in the contour of each breast: a swelling, dimpling of skin, or changes in the nipple.

Then rest your palms on your hips and press down firmly to flex your chest muscles. Left and right breast will not exactly match—few women's breasts do. Again, look for changes and irregularities. Regular inspection shows what is normal for

Not all women are at equal risk for developing breast cancer. Women who are known to be at higher risk include women over 50, women who have a family history of breast cancer, women who experience a late menopause, and women who have never had children. Women with a somewhat higher risk include women who have their first child after 30 and women who began menstruating early.

Women should consult a doctor immediately if they notice a lump or other changes in their breasts, such as a discharge from the nipple. Approximately 80 percent of breast lumps are noncancerous, but the lump needs to be examined by a doctor in order to rule out the possibility of cancer.

Techniques for diagnosis of breast can-

cers include mammography and xerography (both of which involve special X-ray pictures of the breast) and thermography (which uses an infrared scanning device to detect a slight increase in heat that is usually produced by cancer tissue). Mammography is not generally recommended for women under 50, unless they are in certain high-risk groups, because X rays may actually cause breast cancer. Future methods that are being developed for early detection include the T-antigen test, nipple aspiration, and microwave radio signal detection.

## Of the Cervix

Cancer of the cervix is the second most common type of cancer in women. About 2 percent of all women ultimately develop it. It is

## Box 7.5    Continued

you and will give you confidence in your examination.

   **3. Lying down.**   To examine your right breast, put a pillow or folded towel under your right shoulder. Place your right hand behind your head; this distributes breast tissue more evenly on the chest. With the left hand, fingers flat, press gently in small circular motions around an imaginary clock face. Begin at outermost top of your right breast for 12 o'clock, then move to 1 o'clock, and so on around the circle back to 12. (A ridge of firm tissue in the lower curve of each breast is normal.)

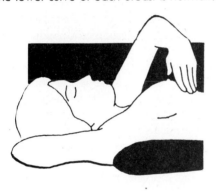

Then move 1 inch inward, toward the nipple and keep circling to examine every part of your breast, including the nipple. This requires at least three more circles. Now slowly repeat the procedure on your left breast with a pillow under your left shoulder and left hand behind your head. Notice how your breast structure feels.

   Finally, squeeze the nipple of each breast gently between the thumb and index finger. Any discharge, clear or bloody, should be reported to your doctor immediately.

[1]From the American Cancer Society. Used by permission.

---

very rare before age 20, but the incidence rises over the next several decades. The average age of women with cancer of the cervix is 45. This disease is more common in women who have had large numbers of sexual contacts and who have borne children.[16] The disease is very rare among Jewish women, which has given rise to the theory that smegma, which tends to collect under the foreskins of uncircumcised males (Jewish males are usually circumcised) may play a part in the development of cancer of the cervix. However, there is no difference in the incidence of cervical cancer among Muslims (males circumcised) and Hindus (males uncircumcised), so the association of cervical cancer and uncircumcised sexual partners is far from established.[17]

   Cancer of the cervix may present no symptoms for five or ten years, and during this period treatment is extremely successful. The well-publicized Pap smear test is the best means now available for identifying cancer of the cervix in the early stages (when it is most susceptible to treatment). This test should be done annually beginning when a woman is 20 (or earlier if a woman is sexually active). From the patient's point of view the Pap smear is an extremely simple test to perform. The physician simply takes a specimen of

---

[16]One study of 13,000 Canadian nuns failed to reveal a single case of cervical cancer. See Novak et al., (1970), p. 212.

[17]Ibid., p. 211.

cervical mucus with a cotton-tipped swab and makes a "smear" of this material on a glass slide. The procedure is quick, simple, and painless. The smear is then stained and examined under a laboratory microscope for the presence of cancerous cells.

As cancer of the cervix begins to invade surrounding tissues, irregular vaginal bleeding or a chronic bloody vaginal discharge develops. Treatment is less successful when the cancer has reached this stage. If treatment (surgery, radiation, or both) is instituted before the cancer spreads beyond the cervix, the five-year survival rate is about 80 percent, but it drops precipitously as the disease reaches other organs in the pelvis. The overall five-year survival rate for cancer of the cervix (including all stages of the disease) is about 58 percent.

One possible cause of cancer of the cervix is the drug DES (diethylstilbestrol, a synthetic estrogen). Between 1940 and 1970, DES was given to many pregnant women who had a history of bleeding, repeated miscarriage, or long periods of infertility. DES was of dubious value in preventing miscarriages, and a greater than usual risk of cancer of the cervix or vagina (a rare cancer) has been found in daughters of women who took the drug while pregnant. It is recommended that women find out whether their mothers took DES and, if so, inform their doctors so that they can be checked for vaginal cancer.

## Of the Endometrium

Cancer of the endometrium (the lining of the uterus) is less common than cancer of the cervix, eventually affecting about 1 percent of women. It usually occurs in women over 35, most commonly in women 50 to 64. Many cases of this cancer are detected by the Pap test, but not all. Thus, in addition to having a regular Pap test each year, women over 35 should report any abnormal bleeding to their doctor in order to rule out the possibility of cancer of the endometrium. The five-year survival rate for endometrial cancer is 77 percent.

Cancer of the endometrium has also been linked to the use of estrogens. Estrogens have proved to be effective in treating certain symptoms of menopause (*see* Chapter 4), but some evidence suggests that the risks of this treatment may outweigh the benefits, especially in postmenopausal women who were given estrogens over a prolonged period of time. The Food and Drug Administration now suggests that if estrogen therapy is employed, it be used cyclically in the lowest effective dose for the shortest possible time with appropriate monitoring for endometrial cancer. There is some recent evidence that the addition of progesterone to the estrogen therapy may decrease the risk of endometrial cancer.

## Of the Prostate

Cancer of the prostate is the second most common form of cancer in men. Nevertheless, cancer of the prostate has never been a significant cause of death, although the mortality rate is rising as the life expectancy has risen. Historically, cancer of the prostate has had a low mortality rate for two reasons: First, it is rare before age 40 and uncommon before age 50. Most cases of prostate cancer occur in males 55 and older. About 25 percent of men in the ninth decade of life have cancer of the prostate. By that time they are likely to die of other causes, such as heart disease, rather than of cancer. Second, most cancers of the prostate are relatively small and grow very slowly. Only a minority spread rapidly to other organs.

Cancer of the prostate, like cancer of the breast, is responsive to sex hormones. Androgens stimulate its growth, and surgical castration (removal of the testicles) is sometimes part of the treatment for this disease. Also, estrogens slow the growth of prostatic cancer and are often given as part of the treatment, again highlighting the physiological antagonism between estrogen and androgen

**Box 7.6**   Chromosomal Abnormalities

In the past decades a series of disorders associated with abnormal sex-chromosome patterns has been identified. A normal female has two X chromosomes (one from the mother and one from the father); and a normal male has one X chromosome (from the mother) and one Y chromosome (from the father) (*see* Chapter 5). But occasionally an individual ends up with an unusual combination of sex chromosomes because of errors that occur before fertilization or immediately after the zygote begins its earliest cell divisions. The most common such anomalies are the XXX (*triple-X syndrome*), XXY (*Klinefelter's syndrome*), XO (*Turner's syndrome*), and XYY patterns.[1]

Triple-X females have been called "superfemales," but are not unusually feminine (or masculine). Because XXX women have been found during chromosome surveys in mental institutions, some people initially concluded that the triple-X syndrome was associated with mental retardation. However, there are many perfectly normal XXX women in the population, and it may be incorrect to assume that XXX women are any more prone to mental problems than other women.

People with Klinefelter's syndrome (XXY) have a relatively normal male body appearance. However, the masculinization is incomplete; and the XXY man has a small penis, small testes, low testosterone production, and therefore incomplete development of secondary sex traits. Some have partial breast development at puberty. XXY men are infertile. These men may have problems with their gender identity, and some exhibit mental retardation.

Females with Turner's syndrome have only one X chromosome (the XO pattern). Ovaries are absent or present only in

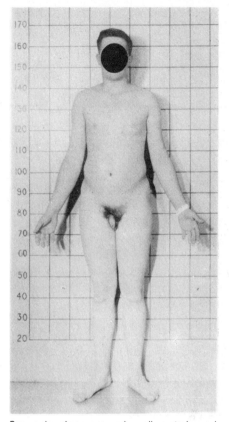

Breast development and small genitals can be seen in this 19-year-old boy with Klinefelter's (XXY) syndrome. From Money, *Sex Errors of the Body* (Baltimore: Johns Hopkins University Press, © 1968), p. 105. Reprinted by permission.

rudimentary form, and if present they do not produce eggs or female hormones. Therefore, the XO female is infertile and does not undergo puberty unless treated with female sex hormones. XO females have short stature. They may have congenital organ defects, webbing between the fingers and toes or between the neck and shoulders. Despite incomplete development of the female organs, XO females are completely feminine in their psychosexual development.

[1]Money (1968), Money and Ehrhardt (1972).

**Box 7.6** Continued

The XYY syndrome originally received a great deal of publicity because a high percentage of such men were found among prisoners who had committed crimes involving violence or sex. XYY individuals are thoroughly masculine in appearance, tend to be more than six feet tall, and show a normal range of intelligence. Does the presence of XYY males in prisons demonstrate a relationship between XYY and aggression? Not if the rate of violent crime of XYY males is no higher than the rate for XY males. A well-designed and well-executed study in Denmark analyzed the sex chromosomes of 4139 men from an unbiased sample and found no higher rate of violent crime among XYY men than among XY men.[2] Thus, the early notion that XYY men are overly aggressive does not appear to be valid.

[2]Witkin et al. (1976).

(compare the use of testosterone in treatment of cancer of the breast). Whereas administration of androgen to women causes beard growth, deepening of the voice, enlargement of the clitoris and may increase sex drive, administration of estrogen to men may cause enlargement of the breasts, impotence, and decreased sexual interest.

The initial symptoms of prostatic cancer are similar to those in benign enlargement of the prostate (which is common in men over 50). They include frequent urination, particularly at night; difficulties initiating urination; and difficulties emptying the bladder. These symptoms largely result from partial obstruction of the urethra by the malignant growth. Early in the course of the disease sexual interest may increase, and frequent erections may occur. Later on, however, there is usually a loss of sexual functioning. A tentative diagnosis of cancer of the prostate can usually be made on the basis of a rectal examination (palpation of the prostate through the rectum), the history of symptoms, and certain laboratory tests. A prostate examination should certainly be part of an annual physical checkup for any man over 50, for, as with other cancers, the prognosis is much more optimistic when it is diagnosed and treated early. The cause of prostatic cancer remains unknown despite efforts to link it with hormonal factors, infectious agents, excessive sexual activity, and sexual frustration.[18]

## Of the Testes

Unlike most other cancers, which strike later in life, cancer of the testes affects young men in the 20–35 age group. In fact, it is the most common cancer in males 29–35. It accounts for only about 1 percent of all cancers in males. Males who have undescended testes or whose testes descended after age 6 are at greater risk for developing testicular cancer. Between 11 and 15 percent of these males will develop cancer of the testes.

If testicular cancer is detected early, it is a very curable cancer. If it is not detected early, it can spread to other parts of the body and cause death. To check for possible symptoms of testicular cancer a male should examine his testes once a month after a warm bath or shower. Using both hands, each testicle is rolled between the thumbs (placed on the top of the testicle) and the index and middle fingers (on the underside). If a hard lump is found it may not be cancerous, but a doctor should be consulted immediately for proper diagnosis. Treatment involves the removal of the affected testicle. A man's sexual activity and fertility are not harmed by removing one

[18]Mostofi (1970), pp. 232–233.

testicle. An implant that is similar to the original testis in size, shape, and feel is inserted into the scrotum.

## Of the Penis

Cancer of the penis is rare in the United States, accounting for about 2 percent of all cancer in males. It is interesting, however, because of its apparent relation to circumcision. Cancer of the penis almost never occurs among Jews, who undergo ritual circumcision within the first two weeks of life, as do most Christians in this country. This disease is also rare, though somewhat less so, among Muslim men, who usually undergo circumcision before puberty. Yet in areas of the world where circumcision is not common, cancer of the penis is much more prevalent. It accounts for about 18 percent of all malignancies in Far Eastern countries, for instance. The usual explanation, though it has not been confirmed, is that circumcision prevents accumulation of potentially carcinogenic secretions, or possibly a virus, around the tip of the penis (the usual site of this type of tumor).

Although this discussion of diseases may leave a few readers with a somewhat unsettled feeling about the potential hazards associated with sexual activity, it surely is not our intention to intimidate. At one time fears of venereal disease and pregnancy were, in fact, emphasized as a means to chastity and virtue, as is expressed in this old limerick:

There was a young lady named Wilde
Who kept herself quite undefiled
    By thinking of Jesus
    And social diseases
And the fear of having a child.

Modern contraceptives and antibiotics like penicillin have removed much of the basis for such fears; some would argue that these products of modern medicine have contributed to a decline in morality by removing the inhibiting force of biological risks associated with sexual activity.

No one to our knowledge is opposed to good health on moral grounds (although Christian Scientists and Jehovah's Witnesses, in the case of blood transfusions, may find certain procedures or the use of medications morally objectionable). Abortions and contraceptive pills or devices, however, raise serious moral questions for many; Roman Catholics in particular. We do not wish to ignore these concerns or treat them lightly and have therefore devoted two entire chapters in our larger textbook, *Fundamentals of Human Sexuality,* to discussions of moral and legal issues related to sex.

Some of the most controversial moral issues today are related to biological and medical developments of the past two decades. Until the advent of reliable, legal, and relatively inexpensive methods of contraception and abortion, moral questions in this area remained somewhat academic for most people. Traditional views which inexorably linked sex with procreation are now being challenged, particularly in light of the threat to the quality of life and even mass starvation that experts ascribe to overpopulation.

Testtube babies are another reality, and this also has led to protests against further research in this aspect of reproductive biology. It is now possible to obtain an egg from a woman's ovary and fertilize it by mixing it with sperm and nutrient fluids in a testtube. The fertilized ovum can then be implanted in a woman's uterus and, in some cases, a normal pregnancy ensues. So far this technique has only been employed in instances where a woman's fallopian tubes were blocked and the egg and sperm were obtained from the woman and her husband. But what if people were to hire someone else to function as a surrogate or host mother on their behalf? Prospective parents could arrange for testtube fertilization of their own germ cells and then arrange (presumably for a fee) to have the fertilized egg implanted in the uterus of the host mother. The host mother would undergo the pregnancy and surrender the child to the biological parents at birth. But what if the parents have divorced in the meantime and refuse to accept the child? Or what if the host mother contracts German measles during the first trimester and the child is born deformed? Is she then liable for the care of the child or are the couple who contracted to produce it responsible? Obviously some understanding of the biological aspects of human sexuality is a necessary but not sufficient condition for providing answers to such questions.

With regard to sexual behavior, an understanding of the anatomy and physiology of the sex organs and related parts of the body can enable one to become more proficient in providing and receiving the pleasure of human sexual contact. But this background alone is not the secret of a joyous sex life any more than a knowledge of engine mechanics and the principles of flight make one a pilot. It is a good beginning, though, and should make the rest much easier.

Abraham, K. *Selected papers of Karl Abraham.* London: Hogarth Press and Institute of Psychoanalysis, 1948.

Arey, L.B. *Developmental anatomy.* 7th ed. Philadelphia: W.B. Saunders Co., 1965.

Austin, C.R., and R.V. Short. *Reproduction in Mammals.* Vol. 1: *Germ cells and fertilization.* Vol. 2: *Embryonic and fetal development.* Vol. 3: *Hormones in reproduction.* Vol. 4: *Reproductive patterns.* Vol. 5: *Artificial control of reproduction.* London: Cambridge University Press, 1972.

Beach, F.A., ed. *Sex and behavior.* New York: John Wiley & Sons, 1965.

Belliveau, F., and L. Richter. *Understanding human sexual inadequacy.* New York: Bantam Books, 1970.

Benedek, T. *Psychosexual functions in women.* New York: Ronald Press, 1952.

Benson, R.C. *Handbook of obstetrics and gynecology.* 3rd ed. Los Altos, Calif.: Lange Medical Publications, 1968.

Berelson, B. Beyond family planning. *Science* 163(1969): 533-543.

Bergstrom, S., M. Bygdeman, B. Samuelsson, and N. Wiqvist. The prostaglandins and human reproduction. *Hospital Practice* 6 (February 1971) 51-57.

Bishop, N. The great Oneida love-in. *American Heritage* (February 1969) 20: 14-17, 86-92.

Borell, U. Contraceptive methods—their safety, efficacy, and acceptability. *Acta Obstet. et Gynecolog. Scand.* 45, Suppl. 1 (1966): 9-45.

Brecher, R., and E. Brecher. *An analysis of human sexual response.* New York: New American Library, 1966.

Calderone, M.S., ed. *Manual of contraceptive practice.* 2nd ed. Baltimore: Williams & Wilkins Co., 1970.

Chertok, L. Psychosomatic methods of preparation for childbirth. *American Journal of Obstetrics and Gynecology* 98(1967): 698-707.

Clark, L. "Is there a difference between a clitoral and a vaginal orgasm?" *Journal of Sex Research* 6, no. 1 (February 1970): 25-28.

Crawley, L.Q., J.L. Malfetti, E.I. Stewart, and N. Vas Dias. *Reproduction, sex, and preparation for marriage.* Englewood Cliffs, N.J.: Prentice-Hall, 1964.

Diamond, M., ed. *Perspectives in reproduction and sexual behavior.* Bloomington: Indiana University Press, 1968.

Dickinson, R.L. *Atlas of human sex anatomy.* 2nd ed. Baltimore: Williams & Wilkins Co., 1949.

Dienhart, C.M. *Basic human anatomy and physiology.* Philadelphia: W.B. Saunders Co., 1967.

Dmowski, W.P., Manuel Luna, and Antonio Scommegna. Hormonal aspects of female sexual response. *Medical Aspects of Human Sexuality,* 8, no. 6 (June 1974): 92-113.

Dodson, A.I., and J.E. Hill. *Synopsis of genitourinary disease.* 7th ed. St. Louis: C.V. Mosby Co., 1962.

Eastman, N.J., and L.M. Hellman. *Williams obstetrics.* 13th ed. New York: Appleton-Century-Crofts, 1966.

Ehrhardt, A.A., and J. Money. Progestin-induced hermaphroditism: I.Q. and psychosexual identity in a study of ten girls. *Journal of Sex Research* 3(1967): 83-100.

Ehrlich, P., and A. Ehrlich. *Population, resources, environment.* 2nd ed. San Francisco: W.H. Freeman & Co., 1972.

Ehrlich, P.R. *The population bomb.* New York: Ballantine Books, 1968.

Ellis, H. *Studies in the psychology of sex.* 2 vols. New York: Random House, 1942. (Originally published in 7 volumes, 1896-1928.)

Fenichel, O. *The psychoanalytic theory of neurosis.* New York: W.W. Norton & Co., 1945.

Ferenczi, S. Male and female: Psychoanalytic reflections on the "theory of genitality," and on secondary and tertiary sex differences. *Psychoanalytic Quarterly* 5(1936): 249-260.

Fisher, S. *The Female Orgasm.* New York: Basic Books, 1973.

Fisher C., *et al.* Cycle of penile erection synchronous with dreaming (REM) sleep. *Archives of General Psychiatry* 12(1965): 29-45.

Fiumara, N.J. Gonococcal pharyngitis. *Medical Aspects of Human Sexuality* 5, no. 5 (May 1971): 195-209.

Ford, C.S., and F.A. Beach. *Patterns of sexual behavior.* New York: Harper & Row, 1951.

Fox, C.A., and B. Fox. Blood pressure and respiratory patterns during human coitus. *Journal of Reproduction and Fertility* 19, no. 3 (August 1969): 405-415.

Fox, C.A., A. Ismail, *et al.* Studies on the relationship between plasma testosterone levels and human sexual activity. *Journal of Endocrinology,* 52(1972): 51-58.

Fox, C.A., H.S. Wolff, and J.A. Baker. Measurement of intra-vaginal and intrauterine pressures during human coitus by radio-telemetry. *Journal of Reproduction and Fertility* 22, no. 1 (June 1970): 243-251.

Freud, S. *The standard edition of the complete psychological works of Sigmund Freud,* James Strachey, ed. London: Hogarth Press and Institute of Psychoanalysis, 1957-1964.

Goodlin, R.C. Routine ultrasonic examinations in obstetrics. *Lancet* (September 11, 1971): 604-605.

Goy, R.W. Organizing effects of androgen on the behavior of Rhesus monkeys. In *Endocrinology and human behavior,* R. Michael, ed. London: Oxford University Press, 1968: 12-31.

Greenblatt, R., E. Jungck, and H. Blum. Endocrinology of sexual behavior, *Medical Aspects of Human Sexuality* 6, no. 1 (January 1972): 110-131.

Greene, F.T. R.M. Kirk, and I.M. Thompson. Retrograde ejaculation. *Medical Aspects of Human Sexuality,* 4, no. 12 (December 1970): 59-65.

Hamburg, D.A., and D.T. Lunde. Sex hormones in the development of sex differences in human behavior. In *The Development of Sex Differences,* E. Maccoby, ed. Stanford, Calif.: Stanford University Press, 1966: 1-24.

Harlow, H.F., J.L. McGaugh, and R.F. Thompson. *Psychology.* San Francisco: Albion Publishing Co., 1971.

Heath, R.G. Pleasure and brain activity in man. *Journal of Nervous and Mental Disease* 154, no. 1 (January 1972): 3-18.

Hewes, G.W. Communication of sexual interest: An anthropological view. *Medical Aspects of Human Sexuality* 7, no. 1 (January 1973): 66-92.

Hollister, L. Popularity of amyl nitrite as sexual stimulant. *Medical Aspects of Human Sexuality,* 8, no. 4 (April 1974): 112.

Jawetz, E., J.L. Melnick, and E.A. Adelberg. *Review of medical microbiology.* 9th ed. Los Altos, Calif.: Lange Medical Publications, 1970.

Jones, H.W., Jr., and W. Scott. *Hermaphroditism, genital anomalies and related endocrine disorders.* Baltimore: Williams & Wilkins Co., 1958.

Kantner, J.F., and M. Zelnik. Contraception and pregnancy: Experience of young unmarried women in the United States. *Family Planning Perspectives,* 5, no. 1 (Winter 1973): 21-35.

————. Sexual experience of young unmarried women in the United States. *Family Planning Perspectives,* no. 4 (October 1972): 9-18.

Karim, S.M.M., and G.M. Filshie. Therapeutic abortion using prostaglandin $F_{2\alpha}$. *Lancet* 1(1970): 157-159.

Kinsey, A.C., W.B. Pomeroy, and C.E. Martin. *Sexual behavior in the human male.* Philadelphia: W.B. Saunders Co., 1948.

Kinsey, A.C., W.B. Pomeroy, C.E. Martin, and P.H. Gebhard. *Sexual behavior in the human female.* Philadelphia: W.B. Saunders Co., 1953.

Knight, R.P. Functional disturbances in the sexual life of women: Frigidity and related disorders. *Bulletin of the Menninger Clinic* 7(1943): 25-35.

Kolodny, R., W. Masters, *et al.* Depression of plasma testosterone levels after chronic intensive marijuana use. *New England Journal of Medicine.* 290, no. 16 (April 18, 1974): 872-874.

Kuchera, L.K. Stilbestol as a "morning-after" pill. *Medical Aspects of Human Sexuality* 6, no. 10 (October 1972): 168-177.

Landtman, G. *The Kiwai Papuans of British New Guinea.* London: Macmillan, 1927.

Laub, D.R., and P. Gandy, eds. Proceedings of the second inter-disciplinary symposium on gender dysphoria syndrome (February 2-4, 1973).

Lloyd, C.W. *Human reproduction and sexual behavior.* Philadelphia: Lea & Febiger, 1964.

Malla, K. *The ananga ranga.* R.F. Burton and F.F. Arbuthnot, trs. New York: G.P. Putnam's Sons, 1964 ed.

Masters, W.H., and V.E. Johnson. *Human sexual inadequacy.* Boston: Little, Brown & Co., 1970.

————. *Human sexual response.* Boston: Little, Brown & Co., 1966.

McClintock, M. Menstrual synchrony and suppression. *Nature.* 229(1971): 244-245.

Millar, J.D. The national venereal disease problem. *Epidemic Venereal Disease: Proceedings of the Second International Symposium on Venereal Disease.* St. Louis: American Social Health Association and Pfizer Laboratories Division, Pfizer, Inc. (1972): 10-13.

Miller, W.B. Sexuality, contraception and pregnancy in a high-school population. *California Medicine,* 119(1973): 14-21.

Mittwoch, U. *Genetics of sex differentiation.* New York: Academic Press, 1973.

Moos, R., D.T. Lunde, *et al.* Fluctuations in symptoms and moods during the menstrual cycle. *Journal of Psychosomatic Research,* 13(1969): 37-44.

Mostofi, F.K. Carcinoma of the prostate. *Modern Trends in Urology,* Sir Eric Riches, ed. New York: Appleton-Century-Crofts, 231-263.

Nefzawi. *The perfumed garden.* R.F. Burton, trs. New York: G.P. Putnam's Sons, 1964 ed. (London: Neville Spearman Ltd., 1963.)

Netter, F.H. *Reproductive system.* The Ciba Collection of Medical Illustrations, Vol. 2. Summit, N.J.: Ciba, 1965.

Neubardt, S., and H. Schulman. *Techniques of abortion.* Boston: Little, Brown, 1972.

Newton, N. Interrelationships between sexual responsiveness, birth, and breast feeding. *Contemporary Sexual Behavior: Critical Issues in the 1970's.* J. Zubin and J. Money, eds. Baltimore: Johns Hopkins University Press, 1973: 77-98.

Noonan, J.T., Jr. *Contraception: A history of its treatment by the Catholic theologians and canonists.* New York: New American Library, 1967.

Novak, E.R., *et al. Novak's Textbook of Gynecology,* 8th ed. Batimore: The Williams & Wilkins Co., 1970.

Patten, B.M. *Human embryology.* 3rd ed. New York: McGraw-Hill, 1968.

Pirages, D., and P. Ehrlich. *Ark II.* San Francisco: W.H. Freeman and Co., 1974.

Pohlman, E.G. *Psychology of birth planning.* Cambridge: Shenkman Publishing Co., 1968.

Pribram, K.H. The neurobehavioral analysis of limbic forebrain mechanisms: Revision and progress report. *Advances in the Study of Behavior,* Vol. 2. New York: Academic Press, 1969: 297-332.

Robertiello, R.C. The "clitoral versus vaginal orgasm" controversy and some of its ramifications. *Journal of Sex Research* 6, no. 4 (November 1970): 307-311.

Rose, R.M. Androgen excretion in stress. *The psychology and physiology of stress,* P.G. Bourne, ed. New York: Academic Press, 1969: 117-147.

Rudel, H.W., F.A. Kincl, and M.R. Henzl. *Birth control: Contraception and abortion.* New York: Macmillan Co., 1973.

Salzman, L. Sexuality in psychoanalytic theory. *Modern psychoanalysis,* J. Marmor, ed. New York: Basic Books, 1968: 123-145.

Schmidt, G., and V. Sigusch. Sex differences in responses to psychosexual stimulation by films and slides. *Journal of Sex Research* 6, no. 4 (November 1970): 268-283.

Schneider, R.A. The sense of smell and human sexuality. *Medical Aspects of Human Sexuality,* 5, no. 5 (May 1971).

Schroeter, A.L. Rectal Gonorrhea. *Epidemic Venereal Disease: Proceedings of the Second International Symposium on Venereal Disease.* St. Louis: American Social Health Association and Pfizer Laboratories Division, Pfizer, Inc. (1972): 30-35.

Sherfey, M.J. The evolution and nature of female sexuality in relation to psychoanalytic theory. *Journal of the American Psychoanalytic Association* 14, no. 1 (1966): 28-128.

Singer, I., and J. Singer. Types of female orgasm. *Journal of Sex Research* 8, no. 11 (November 1972): 255-267.

Thomlinson, R. *Demographic problems: Controversy over population control.* Belmont, Calif.: Dickenson, 1967.

Tjio, J.H., and T.T. Puck. The somatic chromosomes of man. *Proceedings of the National Academy of Sciences* 44(1958): 1222-1237.

Udry, J.R., and N.M. Morris. Distribution of coitus in the menstrual cycle. *Nature* 220(1968): 593-596.

Vatsyayana. *The Kama Sutra.* R.F. Burton and F.F. Arbuthnot, trs. Medallion ed. New York: G.P. Putnam's Sons, 1963.

Wade, N. Bottle-feeding: Adverse effects of a Western technology. *Science,* 184(1974): 45-48.

Wagner, N., and D. Solberg. Pregnancy and sexuality. *Medical Aspects of Human Sexuality.* 8, no. 3 (March 1974): 44-79.

Weiss, H.D. Mechanism of erection. *Medical Aspects of Human Sexuality,* 7, no. 2 (February 1973): 28-40.

Weiss, R.S., and H.L. Joseph. *Syphilis.* Camden, N.J.: Thomas Nelson, 1951.

Westoff, C.F., and L. Bumpass. The revolution in birth control practices of U.S. Roman Catholics. *Science,* 179(1973): 41-44.

Westoff, C.R., and R.R. Rindfuss. Sex preselection in the United States: Some implications. *Science,* 184(1974): 633-636.

Wilkins, L., R. Blizzard, and C. Migeon. *The diagnosis and treatment of endocrine disorders in childhood and adolescence.* Springfield, Ill.: Charles C Thomas, 1965.

Yalom, I.D., *et al.* Postpartum blues syndrome. *Archives of General Psychiatry* 18(1968): 16-27.

————. Prenatal exposure to female hormones: Effect on psychosexual development in boys. *Archives of General Psychiatry,* 28(1973): 554-561.

# Name Index

# Subject Index